CLIMATE CHANGE AND DEVELOPMENT IMPACTS ON GROUNDWATER RESOURCES IN THE NILE DELTA AQUIFER, EGYPT

Marmar Badr Mohamed Ahmed

CLIMATE CHANGE AND DEVELOPMENT IMPACTS ON GROUNDWATER RESOURCES IN THE NILE DELTA AQUIFER, EGYPT

DISSERTATION

Submitted in fulfillment of the requirements of
the Board for Doctorates of Delft University of Technology
and
of the Academic Board of the IHE Delft
Institute for Water Education
for
the Degree of DOCTOR
to be defended in public on
Friday, November 6, 2020 at 12:30 hours
in Delft, the Netherlands

by

Marmar Badr Mohamed AHMED
Master of Engineering in Groundwater Management, Tanta University, Egypt
born in Cairo, Egypt

This dissertation has been approved by the promotors:

Prof. dr. S. Uhlenbrook
Dr. A. Jonoski
Dr. G.H.P. Oude Essink

Composition of the Doctoral Committee:

Rector Magnificus TU Delft	Chairman
Rector IHE Delft	Vice-Chairman
Prof. dr. S. Uhlenbrook	IHE Delft / TU Delft, promotor
Dr. A. Jonoski	IHE Delft, copromotor
Dr. G.H.P. Oude Essink	Utrecht University, copromoter

Independent members:

Prof. dr. ir. L.C. Rietveld	TU Delft
Prof.dr. M.D. Kennedy	IHE Delft / TUDelft
Prof.dr. R. Ludwig	Ludwig-Maximilians-Universität München, Germany
Dr. W.O.A. Soliman	Nile Research Institute, Egypt
Prof.dr. ir. T. Heimovaara	TU Delft, Reserve member

This research was conducted under the auspices of the Graduate School for Socio-Economic and Natural Sciences of the Environment (SENSE)

CRC Press/Balkema is an imprint of the Taylor & Francis Group, an informa business

Published by:
CRC Press/Balkema
Schipholweg 107C, 2316 XC, Leiden, the Netherlands
Pub.NL@taylorandfrancis.com www.crcpress.com – www.taylorandfrancis.com,
ISBN 978-0-367-68345-0 (Taylor & Francis Group)

To my dear parents

I hope this thesis to be a contribution towards your dream

To the one who taught me to talk, read and write

To my mentors in life

To the one who taught me to learn science for the sake of science

I present my humble PhD thesis

To my dear husband

Without you beside me

Without your continuous support

This thesis would have never finished

SUMMARY

Climate change (CC), as predicted by several global climate models, is very likely to have severe impacts in the future, on top of all other global changes. These impacts may have significant influence on natural resources especially surface and groundwater. This influence is particularly problematic for the Mediterranean coastal areas, and especially the northern Nile Delta Aquifer (NDA), where both natural and socio-economic resources of significant values are located. Moreover, population increase and development imperatives create additional pressure on the available water resources. These conditions may eventually lead to insufficient coverage of the needed water demands for agriculture, domestic usage as well as urban and industrial development, unless adaptation and mitigation measures are developed ahead of time.

This thesis has a particular focus on salinization of groundwater resources in the NDA due to saltwater intrusion (SWI). The quality of the groundwater in this area may strongly be affected by the impacts of the sea level rise (SLR), which will lead to increasing salinity concentration in groundwater. In addition, the ongoing and future human activities, especially groundwater extraction will result in the deterioration of the groundwater resources and consequently bring serious negative social and economic impacts. The model was setup based on updated data on groundwater salinization for scattered wells covering the NDA and knowledge of the hydrological, hydrogeological, geological and hydrochemical characteristics of the groundwater in the NDA.

To assess current conditions and develop future adaptation strategies for the NDA, a 3D model simulating regional variable density groundwater flow was implemented, using the SEAWAT code. To identify the representative model for the SWI and salinity conditions in the NDA in the year 2010, a methodology was applied in which the aquifer was evoloved from completely fresh to salinized conditions, using different simulation periods. The model with simulation period that leads to the lowest error in terms of groundwater salinity concentrations was selected as most representative for the NDA. The following nine simulation periods were tested: 200, 400, 600, 700, 800, 900, 1000, 1700 and 2600 years. The results of simulation indicate that the model with simulation period of 800 years captures well the salinity distribution across the NDA. The findings of the simulation indicate that there is a worrisome SWI process in the horizontal and vertical directions in the NDA. Consequently, the groundwater quality is highly deteriorated. Furthermore, the results of the model show that the salinity in the northern area of the NDA could be attributed to SWI. However, the salinity in the southern regions of the NDA is caused by dissolution of minerals from the rocks underlying the aquifer. The model enables the assessment of volumes of groundwater per different types in terms of salinity concentration (fresh, light brackish, brackish and saline), together with their spatial distribution.

The simulated model was applied in NDA for the analysis and prediction of future groundwater resources conditions within pre-defined scenarios of SLR and groundwater extraction. These different scenarios were used for the comparative analysis of the influence of these two factors on the salinization of groundwater resources. Six different scenarios were designed to be tested with the developed model, such that changes in available fresh groundwater can be estimated. Five scenarios were representing conditions in the year 2100, and one was estimating the condition in the year 2500, without any further changes in the system compared to current conditions.

The results from the scenario estimating the condition in the year 2500 is of significance because it indicates that even without any changes in the current conditions, the salinization of the NDA would continue. The results of the other five scenarios show that the potential impact of human interventions such as unplanned groundwater extraction is far more significant for the NDA than the expected SLR impact. The model also allows for more detailed analysis of fresh groundwater availability in different governorates.

Given the results from the scenario analysis, which indicate that the fresh groundwater availability will most likely continue to decrease in the NDA, (potentially up to 20 %); there is a need to investigate possible adaptation measures. Even though the results presented here are mostly about the NDA as a whole, the overall conditions vary significantly across this large area. The situation is different in some governorates that are characterized by larger irrigation areas and larger population density. Consequently, there are differences in groundwater extraction rates. These differences may influence the possibility of implementing particular adaptation measures, and therefore their eventual selection.

The developed model was used for testing the implementation of adaptation measures for one of the most affected governorates - Sharkeya. The Sharkeya governorate has been selected because it is characterized with quite extensive groundwater extraction rate (682 x 10^6 m^3/year), a very large cultivated irrigated area (ranked the third in the country in terms of crop production) and large variations of salinity concentrations of groundwater (21 kg/m^3 - 0.2 kg/m^3).

Three different adaptation measures have been selected for testing in the Sharkeya governorate, well injection using tertiary treated wastewater, extraction of brackish groundwater, and changing of cropping patterns and irrigation practices. The three adaptation measures are assumed to be implemented during the same period of analysis used for the scenarios (till year 2100), and their effectiveness was assessed using gained/lost fresh groundwater volumes. The results indicate that in terms of increase of fresh groundwater inside the aquifer the best results can be expected from well injection. Regarding extraction of brackish groundwater, it can be said that it does not bring significant reduction of fresh groundwater volumes in the aquifer in the period till 2100. After desalination, this measures provides additional $9x10^9$ m^3 of ready-to-use fresh water over the same period (2010-2100). Changing of cropping patterns and irrigation practices also do not bring significant changes inside the aquifer, although the amount of water saved overall can be quite significant. The implementation of this measure on large scale, however, is a challenging process that may take long period of time and resources (investments, changes of policies, training of farmers etc.). These considerations, together with other implementation-related aspects, need to be taken into account for the final choice of (combined) adaptation measures.

SAMENVATTING

Klimaatverandering (CC), zoals voorspeld door verschillende mondiale klimaatmodellen, zal in de toekomst waarschijnlijk ernstige effecten hebben, bovenop alle andere wereldwijde veranderingen. Deze effecten kunnen een aanzienlijke invloed hebben op natuurlijke hulpbronnen, met name oppervlakte- en grondwater. Deze invloed is met name problematisch voor de kustgebieden van de Middellandse Zee, en in het bijzonder de noordelijke Nijldelta-aquifer (NDA), waar zowel natuurlijke als sociaal-economische hulpbronnen van grote waarde zijn gevestigd. Bovendien zorgen bevolkingsgroei en ontwikkelingsverplichtingen voor extra druk op de beschikbare watervoorraden. Deze omstandigheden kunnen uiteindelijk leiden tot onvoldoende dekking van de benodigde waterbehoeften voor landbouw, huishoudelijk gebruik en stedelijke en industriële ontwikkeling, tenzij van tevoren aanpassingen en mitigatiemaatregelen worden ontwikkeld.

Dit proefschrift richt zich in het bijzonder op verzilting van grondwaterbronnen in de NDA als gevolg van zoutwaterintrusie (SWI). De kwaliteit van het grondwater in dit gebied kan sterk worden beïnvloed door de effecten van de zeespiegelstijging (SLR), wat zal leiden tot een verhoogde zoutconcentratie in het grondwater. Bovendien zullen de voortdurende en toekomstige menselijke activiteiten, voornamelijk de winning van grondwater, leiden tot een verslechtering van de grondwatervoorraden en ernstige negatieve sociale en economische effecten veroorzaken. Het model is opgesteld op basis van bijgewerkte data over verzilting van grondwater voor verspreide putten die de NDA bestrijken en kennis van de hydrologische, hydrogeologische, geologische en hydrochemische kenmerken van het grondwater in de NDA.

Om de huidige omstandigheden te beoordelen en toekomstige adaptatiestrategieën voor de NDA te ontwikkelen, is een 3D-model geïmplementeerd dat regionale grondwaterstroming met variabele dichtheid simuleert, gebruikmakend van de SEAWAT-code. Om het representatieve model voor de SWI en saliniteitsomstandigheden in de NDA in 2010 te identificeren, is een methodologie toegepast, waarbij de toestand in de aquifer evolueert van volledig zoet tot verzilt water, gebruikmakend van verschillende simulatieperioden. Het model met simulatieperiode dat leidt tot de kleinste foutmarge met betrekking tot het zoutgehalte in het grondwater, is geselecteerd als meest representatief voor de NDA. De volgende negen simulatieperiodes zijn getest: 200, 400, 600, 700, 800, 900, 1000, 1700 en 2600 jaar. De resultaten van de simulatie geven aan dat het model met een simulatieperiode van 800 jaar de verdeling van het zoutgehalte over de NDA goed vastlegt. De bevindingen van de simulatie geven aan dat er in de NDA een zorgwekkend SWI-proces is in horizontale en verticale richting. Hierdoor wordt de grondwaterkwaliteit sterk verslechterd. Bovendien laten de resultaten van het model zien dat het zoutgehalte in het noordelijke deel van de NDA aan SWI kan worden toegeschreven. Het zoutgehalte in de zuidelijke regio's van de NDA wordt echter veroorzaakt door het oplossen van mineralen in de onderliggende rotsen. Het model bemogelijkt het bepalen van het volume van grondwater per type zoutconcentratie (zoet, licht brak, brak en zout), inclusief de bijbehorende ruimtelijke distributie.

Het gesimuleerde model is in de NDA toegepast voor de analyse en voorspelling van toekomstige toestand van grondwatervoorraden binnen vooraf gedefinieerde scenario's van SLR en grondwaterwinning. Door middel van de scenario's is de invloed van deze twee factoren op de verzilting van grondwatervoorraden met elkaar vergeleken. Er zijn zes verschillende scenario's ontworpen om te testen met het ontwikkelde model, zodat veranderingen in beschikbaar zoet grondwater kunnen worden geschat. Vijf scenario's vertegenwoordigden de omstandigheden in het jaar 2100 en één scenario schatte de toestand in het jaar 2500, zonder verdere wijzigingen in het systeem in verhouding tot de huidige omstandigheden.

De resultaten van het scenario dat de toestand in het jaar 2500 schat, zijn van belang omdat het aangeeft dat de verzilting van de NDA zelfs zonder veranderingen in de huidige omstandigheden zou doorgaan. De resultaten van de andere vijf scenario's laten zien dat de potentiële impact van menselijke ingrepen zoals ongeplande grondwaterwinning voor de NDA veel significanter is dan de verwachte SLR-impact. Het model bemogelijkt ook een gedetailleerdere analyse van de beschikbaarheid van zoet grondwater in verschillende provincies.

Gezien de resultaten van de scenarioanalyse, die erop wijzen dat de beschikbaarheid van zoet grondwater in de NDA hoogstwaarschijnlijk zal blijven afnemen (mogelijk tot 20%), is het nodig mogelijke aanpassingsmaatregelen te onderzoeken. Hoewel de hier gepresenteerde resultaten meestal over de NDA als geheel gaan, variëren de algemene omstandigheden aanzienlijk over dit grote gebied. De situatie is anders in sommige provincies, die worden gekenmerkt door grotere irrigatiegebieden en een grotere bevolkingsdichtheid. Ten gevolge hiervan, zijn er verschillen in de winningspercentages van grondwater. Deze verschillen kunnen van invloed zijn op de mogelijkheid om een bepaalde aanpassingsmaatregel te implementeren en dus op hun uiteindelijke selectie.

Het ontwikkelde model is gebruikt voor het testen van de implementatie van aanpassingsmaatregelen voor een van de meest getroffen provincies - Sharkeya. Sharkeya is geselecteerd omdat het wordt gekenmerkt door een vrij significante grondwaterwinning (682 x 10^6 m^3 / jaar), een zeer groot geïrrigeerd gebied (derde in het land in termen van gewasproductie) en grote variaties in zoutconcentraties van grondwater (21 kg / m^3 - 0, 2 kg / m^3).

Er zijn drie verschillende aanpassingsmaatregelen getest in de Sharkeya provincie, putinjectie met tertiair behandeld afvalwater, extractie van brak grondwater en de wijziging van teeltpatronen en irrigatiepraktijken. De drie aanpassingsmaatregelen worden verondersteld te worden uitgevoerd in dezelfde analyseperiode die voor de scenario's is gebruikt (tot 2100) en hun effectiviteit is bepaald aan de hand van verloren / gewonnen hoeveelheden zoet grondwater. De resultaten geven aan dat in termen van toename van zoet grondwater in de aquifer de beste resultaten te verwachten zijn bij het gebruik van putinjectie. Met betrekking tot de extractie van brak water kan worden gesteld dat deze in de periode tot 2100 geen significante vermindering van de hoeveelheden zoet grondwater in de aquifer met zich meebrengt. Na ontzilting levert deze maatregel in dezelfde periode (2010-2100) 9-x10^9 m^3 zoet water, dat klaar voor gebruik is. De verandering van teeltpatronen en irrigatiepraktijken brengen ook geen significante veranderingen in de aquifer, hoewel de hoeveilheid water dat bespaard wordt significant kan zijn. Echter, de implementatie van deze maatregel op grote schaal is een uitdagend proces dat veel tijd en middelen (investeringen, beleidswijzigingen, opleiding van boeren, enz.) kan vergen. Deze overwegingen, samen met andere uitvoeringsaspecten, moeten in aanmerking worden genomen bij de uiteindelijke keuze van (gecombineerde) aanpassingsmaatregelen.

ACKNOWLEDGMENTS

This thesis will have never finished without Dr. Andreja and Dr. Gu.

I recall the long corridor with white karara marble flooring as I was always looking to the floor feeling guilty after long wasted winter in my country with zero progress, knocking on Dr. Andreja's office. Then, I meet this supportive understanding face giving me hope and another chance to continue and restart again. Understanding challenges and obstacles that I face in my hometown, appreciating family commitment and organizing with me another work plan to adapt the situation. He keeps on giving me lessons not only in numerical modeling but also in humanity. No words can describe my gratitude to you Andreja and I consider myself very fortunate to have you as my supervisor. You taught me how to simplify complicated issues in a very detailed and logical consequence. Thank you for everything, for unlimited support and motivation, for your continuous open door, for your patience, for your guidance, and above all, for always being there for me with your valuable effort and time in spite how busy you are. We have passed together a very long hard route. I have never felt alone. Even in the darkest time when my laptop was stolen with all the data and 3 years of work, you were there supporting me and encouraging me. Thank you for being the best supervisor ever.

Gu, in your first lesson in groundwater modeling, while teaching SWI and commenting on a figure that contains all the deltas around the world except Egypt (it was missing in the figure), you said that you are ready to supervise any research on all deltas around the world. This was our first meeting and I was very happy to be under your guidance and supervision since then. I was fascinated with your experience in the topic and every meeting I learned more and more from your side. You have helped me a lot in spite of being a very critical reviewer. I must say that sometimes your comments were frustrating. I remember that in one meeting your comments made me work for 6 month back in the model but by the end, I realized it was all for my good and I appreciate your attention to perfection. I will never forget the third paper and its final review before publishing (it was already accepted). You have worked with me until 2 am to refine it. Around 15 mails on that night, step by step, you had revised, refined and even prepared some figures until being satisfied. Whenever you were in Cairo, you give me long fruitful meetings in spite how busy you were. In addition, of course to our meetings together with Andreja in Delft that could extends to 9 hours for final push to wrap up everything especially during the last months of my Ph.D. From my heart, thank you.

I would like to thank Prof. Stefan Uhlenbrook for accepting me as a Ph.D. fellow in IHE that opened for me a new horizon of education. Thank you for your patience and guidance.I know in spite of your busy schedule you are always there when needed.

I am very grateful to many friends who made me feel family away from home, Shakeel my dear Pakistani friend and his lovely family. Thank you for my secret palace that taught me very important lessons in my life. Among them, that when you close your eyes there is no difference between a humble clean place and a very luxurious one, if you have peace of mind. Thanks Syria for your lovely delicious dinners after a long tiring day in IHE. Thanks Hamed for being there always for me. Thank you.

Throughout my Ph.D. journey, I had the luck to get in touch with some unique friends with different personalities, perspectives, cultures and background. Each one touches my life differently and had a big impact on me. Thank you, Zahraa and Reem, (my sisters) for always being there for me in the most critical situations. Patricia my sincere gratitude goes to you for saving Omars' life and me in the first couple of months in IHE. Without knowing me you have given me the first welcome in a very hard time. Heba and Amer, thank you for what you have done with Youssef during my stay in the hospital. Tonneka, thanks for being so kind and friendly to me. I remember your smile when you see me and advise that family is the most important thing. Thanks Gordon de Wit for giving me life again after you retrieved all the data on my laptop after being completely erased when it was stolen. It was great to know such diversity in cultures, people and cuisines within my Ph.D. path. Thanks to restaurant stuff and their continuous smile and delicious food, that made my life easy and thanks to reception stuff for their welcoming spirit.

Yasmin el Nemr, Shaimaa, Elham, Jakia, Sondos, Ahmed Farrag, Shahnor, Gerda, Marielle, Maria, Jolanda, Mosad, Yassir aly, Eman Fadel, Ebdy, Yomna, Ahmed ghandour, Ahmed Ragab, Hisham, Taha, and Abeer. Thank you all for your support and life sharing moments that I will never forget.

Aunt Botheyna and uncle Medhat my father in law, May GOD bless his soul, you both had supported me a lot throughout my travel to Delft. I would like to thank you for your unconditional love, care and peaceful spirit. Thanks Somaya, Azza and Aya, my dear sisters for your support and love.

My dear husband, you are the one who have supported me all this hard journey and I know that our home had suffered a lot because of this. You have always encouraged me to continue even in the darkest time. You taught me that I can do it and that any obstacle I face with consistency and persistence will be conquered. You shared with me disappointment and hopes and above all, you gave me the push to finalize. Finally, but at the top of everything, my sons, Youssef and Omar. They spent a large part of their childhood in a fragmented house. I love you and I wish to you happiness in your lives and that one day you come to the Netherlands and study.

CONTENTS

1 INTRODUCTION

1.1 INTRODUCTION TO THE STUDY AREA

Egypt lies in the northeastern corner of Africa with a total area of about one million km^2. It consists of a vast desert plateau crossed by the Nile Valley and the Nile Delta (ND) that represents about 5 percent of the Egyptian area. The majority of Egypt's surface area is desert. Most of the agricultural land lies close to the banks of the River Nile and its Delta [1].

The ND is the food basket of Egypt. It is the most fertile land where about 60% of Egypt's population lives. Like most of the deltas around the world, agriculture activities are dominant in the ND due to the nature of the soil and the presence of an irrigation system. Consequently, it has a great economical and residential importance to Egypt (Figure 1.1).

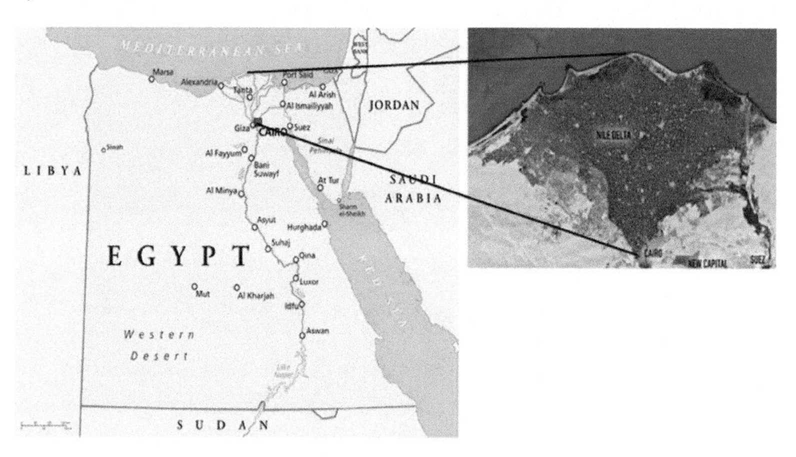

Figure 1.1. The location map of the Nile Delta

The Egyptian population has increased from 22 million in 1950 to around 100 million in 2019. It is even predicted that the population will increase to more than 120 million by 2050 [2]. This rapid increase of population decreases the water resources allocated per capita. Currently, the water allocation per capita in Egypt is 720 m^3/year [2]. If the population growth continues with this trend, the water allocated per capita will severely decline to critical levels. Over the coming years, this declining trend will cause serious risks.

Nile water alone is no longer sufficient for the increasing water requirements of different development activities in Egypt and the highly increasing population. Therefore, groundwater is increasingly being exploited. Extensive extraction of fresh water from the Nile Delta Aquifer (NDA) intensifies saltwater intrusion (SWI) [3]. An emerging problem has a huge impact on the salinization of wells leading to several undesirable consequences. Moreover, this valuable resource is being threatened by SWI due to sea level rise (SLR), a common problem for all coastal aquifers around the world [4].

SLR is one of the expected global warming impacts due to climate change (CC) [5]. The Egyptian coastal area of the ND is one of the highly sea level rise (SLR)-vulnerable regions in the world due to its low elevation [5]. The SLR accelerates the SWI into the NDA and hence affecting the quality of the groundwater. It is predicted that SLR will directly affect more than 3.8 million capita and an area of about 1800 km^2 will be submerged in the ND [6]. This calls for a rapid adaptation and management plan.

The issue of the management of groundwater in the ND has become one of the top priorities in the Egyptian water agenda, to meet not only the economic needs but also the social, cultural and environmental needs. While there is almost a consensus that CC has currently already induced changes in groundwater salinization, it is projected that they will amplify in the future [5]. There is a gap of knowledge about the deterioration in groundwater salinization, a situation which is resulting from lack of suitable hydrological monitoring and modeling systems. In particular, current projections and adaptation measures for future salinization scenarios are very limited. This stresses the need to understand and analyze the hydrological conditions and their impact on groundwater behavior in the NDA, in addition to studying the negative impacts of SLR and development on the surrounded environment, from a new perspective focusing on groundwater quality. This could be achieved by simulating the current condition and predicting the future scenarios with reliable updated hydrological data series.

A concrete adaptation measure could be then built on the output of those realistic simulations. The approach taken in this study can be applicable for other coastal aquifers. Whereas Nile Delta is no exception. In spite of differences in geometry and their hydrological data, most deltaic areas face similar development and climate stresses.

1.2 OBJECTIVES AND RESEARCH QUESTIONS

The overall goal of this research is to contribute towards ensuring quality and sustainability of groundwater in the NDA as a strategic and economic source for life and development. Sustainable groundwater resources development and environmentally sound protection should be an end goal. Their attainment is closely linked to water resources planning and management and influenced by economic and social constraints.

Note that a review of the state of the knowledge and related knowledge gaps are presented in details in chapter 2.

1.2.1 Main objective

The broad objective of this research is to contribute to the development of a framework for the long term planning for exploitation and sustainable management of groundwater resources in the NDA. One main contribution to the framework is the developed variable-density 3D groundwater model that will be used to address the future groundwater resources development within several scenarios of CC impact (SLR) and the impact of development (groundwater extraction).

1.2.2 Research questions

1. What is the current knowledge regarding groundwater salinization in the study area, and where are the knowledge gaps?
2. What is the current situation of salinization in the Nile Delta and its governorates? What are the recommended locations for extraction from the current perspective (2010)?
3. What is the impact of saltwater intrusion under the various proposed future scenarios of climate change (sea level rise) and development (groundwater extraction) in the whole Nile Delta Aquifer?
4. What are the best locations and the vulnerable ones for groundwater extraction in the Nile Delta governorates in a long-term perspective?
5. What are the proposed adaptation measures that minimize the loss of fresh groundwater due to saltwater intrusion, and what are their limitations?

1.3 METHODOLOGY

This research addresses the impact of SLR and excessive groundwater extraction and their consequences on the NDA through a number of steps. The following methodological steps have been followed.

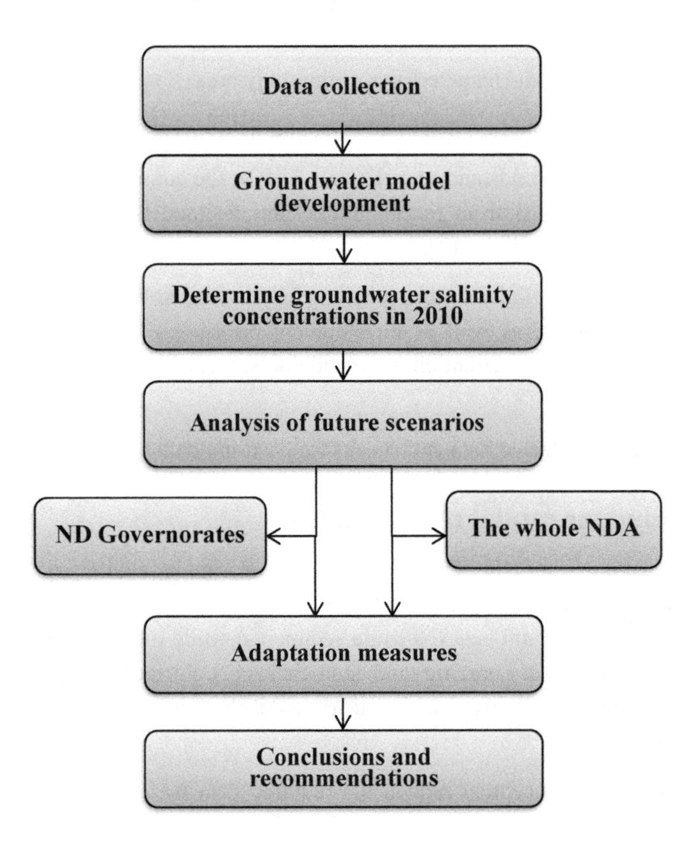

Figure 1.2. Flow chart of the methodological steps

This chapter provides a general overview over the followed methodology; further details of methods applied in each study are given in the respective chapters.

1.3.1 Data collection

For better management of groundwater resources, it is crucial to have enough reliable data about the physical, hydrological and hydro-(geo)logical settings of the study area. Physical settings include land use, meteorological data, topography and soil classification. The data was collected from different private and governmental entities. This phase included the following stream of activities:

1. Identifying the general characteristics of the study area.
2. Collecting previous studies, historical data, and field investigations.
3. Reviewing climate trends and scenarios for SLR at the country level based upon examination of results from recent global climatic models, as well as results published in scientific literature.
4. Reviewing current salinity control projects that are funded by the government or international donors, as well as development priorities of Egypt, in order to determine the degree of attention to potential risks posed by CC on groundwater sector.
5. Analysis of the collected data and checks for ensuring data quality.

1.3.2 Groundwater model development

Conceptually speaking, the ND region consists of two main aquifer systems, the Holocene and the Pleistocene aquifers. The Holocene aquifer is composed of medium to fine-grained sand, silt, clay and peat (Newnile sediments) while the Pleistocene aquifer is composed of thick layers of quarzitic sand and pebbles [7]. Surface irrigation network and main branches of the Nile were included to model the exchange between surface water and groundwater.

In order to develop the conceptual model, the required data was collected, organized and digitized as follows:

- Contour maps showing the elevation and thickness of different aquifer layers.
- Geological and cross sections maps of the NDA.
- Contour maps of porosity, hydraulic conductivity, groundwater heads, salinity concentrations in different depths, canals and drains network data.
- Groundwater extraction and observation wells' data.

The system was simulated by using the MODFLOW and SEAWAT codes to determine the spatial distribution of salinity concentrations in the NDA. The proposed model combination fits best for the research because it describes the physical processes of SWI well, it provides wide range of possibilities to simulate groundwater management and

SLR related scenarios, the model codes have been widely used in the world and most of the required data are available within the time and resource limits of this study.

In the research, the initial salinity concentration distribution of the model of the NDA is completely fresh with only saline concentration at the Mediterranean seaside. Our target is solely to determine the appropriate simulation period, which provides the best match between the modeled and observed salinity concentration data as, will be discussed in details in chapter 3. The final goal is to have a regional 3D groundwater salinity model that confidently represents the present situation (2010) which can be used for future predictions. We designated the year 2010 as the reference year, as most data are available for this year.

1.3.3 Analysis of future scenarios

The final developed model results are used as initial conditions in order to simulate future conditions under several proposed scenarios. These scenarios are prepared to cover different aspects of SLR and groundwater extraction for the period until 2100. Extreme conditions are examined together, e.g. high extraction levels and high SLR to determine the impacts, especially in terms of distribution of different types of groundwater with regards to salinity concentration and focusing on freshwater availability. Also, low extraction levels with high SLR are examined in order to make a comparative analysis of SLR impact on groundwater salinization versus human interventions. An assessment for the whole NDA and for individual governorates is carried out with the outputs from the scenarios proposed.

1.3.4 Adaptation measures

The final step is devoted to the proposal of adaptation measures and solutions based on the model outputs and the overall research work. There are different possible solutions to control SWI where each measure is studied thoroughly. Three adaptation measures are discussed in one of the ND governorates (Sharkeya governorate). The three measures are: well injection, extraction of brackish water and changing of cropping patterns and irrigation practices Groundwater salinization conditions are assessed with and without adaptation measures, using model simulations. Each measure of adaptation is analysed from the obtained results and other considerations. The advantages and disadvantages of different adaptation measures are discussed in details in chapter 5.

1.4 STRUCTURE OF THE THESIS

The thesis consists of six chapters as follows:

Chapter 1: Presents the general outline of the thesis. It starts introducing the study area and the problems it faces. The chapter then presents the main objective of the research, research questions and the methodology undertaken.

Chapter 2: Gives an overview of the previous studies of groundwater management and SWI modeling in the NDA in Egypt and worldwide.

Chapter 3: Presents the development of the groundwater simulation model for salinity distribution in the NDA and the obtained results for the current conditions (in year 2010).

Chapter 4: Demonstrates the simulation of salinity distribution for the proposed scenarios for the years 2100 and 2500 in the NDA as a whole, and in the ND governorates. Comparative analysis between the impact of groundwater extraction and/or SLR on salinization is also presented.

Chapter 5: Presents three adaptation measures to deal with future salinization for one selected ND governorate.

Chapter 6: Finally, this chapter provides conclusions and recommendations for future research.

References

1. EGSA., Egyptian General Survey and Mining: topographical map cover ND, scale 1: 2 000 000; Egyptian General Survey and Mining publishing centre, Cairo, Egypt, 1997.

2. CAPMAS, Central Agency for Public Mobilization and Statistics Egypt (www.capmas.gov.eg), 2020.

3. Morsy, W. S., Environmental management to groundwater resources for ND region, Ph.D. thesis, Fac. of Eng., Cairo Univ, Egypt, 2009.

4. Werner, A.D., Bakker, M., Post, V.E.A., Vandenbohede, A., Lu., C., Ataie-Ashtiani, B., Simmons, C.T., Barry, D.A. Seawater intrusion processes, investigation and management: recent advances and future challenges. *Adv. Water Res. J.* 51, 3-26, 2013.

5. Oppenheimer, M., B.C. Glavovic, J. Hinkel, R. van de Wal, A.K. Magnan, A. Abd-Elgawad, R. Cai, M. Cifuentes-Jara, R.M. DeConto, T. Ghosh, J. Hay, F. Isla, B. Marzeion, B. Meyssignac, and Z. Sebesvari: Sea level rise and implications for low-lying islands, coasts and communities. In: IPCC special report on the ocean and cryosphere in a changing climate, 2019.

6. El Raey, M., Fouda, Y., and Nasr, S., GIS assessment of the vulnerability of the Rosetta area, Egypt to impacts of sea rise, *Environ. Monitor. Assess. J.* 47, 59–77, 1997.

7. Saleh, M. F., Some hydrological and hydrochemical studies on the ND, MSc. thesis, Fac. of Sci., Ain Shams Univ, Egypt, 1980.

2 A REVIEW OF SEAWATER INTRUSION IN THE NILE DELTA GROUNDWATER SYSTEM

The content of this chapter is based on the published article:

Mabrouk, M.; Jonoski, A.; Oude Essink, G.H.P.; Uhlenbrook, S. A Review of seawater intrusion in the Nile Delta groundwater system - The basis for assessing impacts due to climate changes, SLR and water resources development, *Nile Water and Eng. J.* 1, 46-51,2017

2.1 ABSTRACT

Serious environmental problems are emerging in the River Nile basin and its groundwater resources. Recent years have brought scientific evidence of climate change (CC) and development-induced environmental impacts globally as well as over Egypt. Some impacts are subtle, like decline of the Nile River water levels, others are dramatic like the salinization of the coastal aquifer of the Nile Delta (ND) - the agricultural engine of Egypt. These consequences have become a striking reality causing a set of interconnected groundwater management problems. Massive population increase that overwhelmed the ND region has amplified the problem. Many researchers have studied these problems from different perspectives using various methodologies and objectives. However, the researchers all confirmed that significant groundwater salinization has affected the ND and this is likely to become worse rapidly in the future. This chapter presents, categorizes, critically analyses and synthesizes the most relevant research regarding CC and development challenges in relation to groundwater resources in the ND. It is shown that there is a gap in studies that focus on sustainable groundwater resources development & environmentally sound protection as an integrated regional process in the ND. Moreover, there is also a knowledge gap related to the salinization deterioration of groundwater quality. The chapter recommends further research that covers the groundwater resources and salinization in the whole ND based on integrated 3D groundwater modeling of the Nile Delta Aquifer (NDA).

2.2 INTRODUCTION

Among all current environmental and social changes, CC will have severe future impacts in delta areas [1]. There is a wide range of impacts including: sea level rise (SLR), changes in rainfall patterns, floods and droughts frequencies, salinization levels, and settlement of land. These impacts may have significant influence on natural resources, especially water resources - either surface water or groundwater. This is particularly problematic for the Mediterranean coastal areas, and especially the northern ND coast in Egypt [2].

The ND in Egypt is occupied by the most populated governorates in Egypt. About 60% of Egypt's population lives in the ND region [3, 4]. Agriculture activities are predominant in the region (around 63% of the total agricultural land of Egypt) due to the nature of the soil and the irrigation system [4]. The NDA is a vast leaky aquifer that is located between Cairo and the Mediterranean Sea [5]. The productive aquifer is bound by an upper semi-permeable layer and lower impermeable rocky layer [5]. The aquifer is recharged by infiltration from excess irrigation water and the very limited rainfall that infiltrates through the upper clay layer [6].

The quality of the groundwater in this area may be strongly affected by the impact of SLR combined with changes of Nile River flows, leading to an increase in the salinity levels of groundwater [7]. In addition, the current and future human activities, especially extensive and unplanned groundwater extraction, are resulting in deterioration of the available groundwater resources [8]. Serious negative socioeconomic impacts can follow as a consequence [9]. This situation prompts for studying and analyzing the problem thoroughly and identifying flexible adaptation strategies that can not only mitigate the negative effects of CC, but also lead to capacity development for coping with uncertain future changes.

Many water researchers have been interested in the ND, and their studies tackled it from different aspects, focusing on either surface water or groundwater. Different tools have been used to characterize, classify and analyze the groundwater aquifer. Most of the studies agreed that CC is a significant issue that should be considered with high priority [1]. A number of researchers investigated the problem of current water quality status of groundwater, but few studies cover the whole ND e.g. [10]. Also, most of the strategies for adaptation measures focus only on a limited area and do not take into consideration the combined effects that may become apparent when studying the ND from a regional perspective.

This chapter attempts to identify and analyze the findings of most recent studies regarding CC and development challenges that the ND faces with particular focus on its groundwater resources. This analysis should serve as the basis for identifying future research needs. As will be demonstrated, the main drawback of existing research efforts is their local focus, leading to the need for an integrated approach that takes the whole ND as a unit for analysis. Furthermore, this chapter proposes research needs for such approach that should lead to sustainable solutions. The proposed approach focuses mainly on different hydrological, hydro-(geo)logical, geological and hydro-chemical characteristics of the groundwater aquifers in the ND and incorporates them in a 3D groundwater model that can serve as one of the predictive tools for analyzing possible future sustainable solutions.

The structure of the chapter is as follows: Section 2 provides an overview of the studies related to CC impacts, particularly SLR, on the ND. Section 3 introduces the NDA and an overview of the research studies related to identifying its hydro-(geo)logical, hydrological and salinity characteristics. Existing modeling approaches with SEAWAT and specific groundwater modeling studies of the ND are introduced in section 4, followed in section 5 by an overview of studies related to possible adaptation and mitigation measures. In section 6, the identified knowledge gaps are discussed. The chapter ends with a section that proposes further research directions for assessing CC and development-related impacts on the groundwater resources of the NDA.

2.3 CLIMATE CHANGES AND NILE RIVER

Understanding CC implications in the Nile basin has attracted many researchers worldwide. The first impact considered is related to potential changes in precipitation and temperature patterns that may lead to changes in the Nile flows [11]. [12] have combined six climate models with an aggregated monthly water balance model that use precipitation fields generated from the climate models. The results of their research that covers the whole Nile Basin indicated that five of the climate models predicted an increase in Nile flow at Aswan. On the other hand, [13] studied the Nile flow patterns using nine representative samples from the full range of CC scenarios. Using water balance models, the results of eight out of nine scenarios in that research showed a high tendency for a decrease in Nile flows. [11] discussed a number of studies that dealt with future CC in the Nile Basin and the recent models applied. The authors highlighted that the studies of CC and its influence on flow patterns over the Nile Basin provide conflicting evidence for long term trends. Although, there is no significant change regarding the overall pattern of flow or precipitation, the trends (increase / decrease) are highly uncertain. The authors therefore emphasized the importance of further CC impact studies.

Another significant impact of CC is SLR [1]. Egypt is considered among the most vulnerable countries, according to [14] and [1]. Fluctuations in MSL will affect delta regions causing saltwater intrusion (SWI) and shoreline retreat [15]. [16] studied 33 deltas around the world. Their studies found that approximately 85% of the deltas worldwide experienced flooding which results in temporary submergence. They concluded that the vulnerability to flooding in delta regions around the world could increase by 50% under the projected values for SLR in the 21 century. Their studies attributed the reason behind the sinking of deltas to human activities due to removal of oil, gas and water in addition to SLR.

The SLR along the Egyptian coast has been studied by many scientists. [17, 18] used bio-sedimentological indicators and tide gauge data for SLR estimation. [19, 20] used different climate models to predict SLR. The range of SLR predicted for the coming 100 years, lies between 30 and 150 cm along the Mediterranean Sea. The most common estimate that is repeated in many reviews is 60 cm [15]. [21] examined the SLR in three coastal cities, Alexandria, Portsaid and Suez, using five different statistical models: linear, quadratic, logarithmic, exponential and power models. Their results show that the SLR is not uniform in the three cities. In Alexandria, the annual rate ranges between 1.94 and 2.22 mm/year, in Port Said, it is between 2.74 and 3.57 mm/year and in Suez on the Red Sea, it ranges between 0.90 and 1.94 mm/year. It should be mentioned that some other studies showed different future SLR and SWI in the coastal zone of the ND e.g. [22, 14, 23, 24, 25, 26, 27]. [14] predicted that the increase in SLR in the coastal region of the ND will lead to flooding in the eastern region and a severe damage to harbors. [24] predicted that a 49 cm SLR by the year 2050 is likely to cause salinization in the river mouth of 500-800 mg/l. [25] studied the economic and social impact that could be induced due to SLR. Their studies found that the SLR will lead to the loss of a large area of touristic villages and harbors that have great economic value to Egypt, even more than agriculture. These studies were based on less reliable data and assumed that SLR would be linear in time. However, according to [21], SLR is expected to accelerate as a function of time.

There are different studies worldwide that have compared between the impact of extraction and SLR on SWI e.g. [28]. However, limited studies made the assessment whether SLR is the only responsible factor for increased SWI in the ND or not. Extensive groundwater extraction is also a very significant factor that increases SWI in the ND [29]. [30] added that the recycling of sewage water have engendered soil salinization in the northern ND. Groundwater wells which were beyond salinization zones in the past are consequently showing up-coning of saline or brackish water [31]. It is in fact considered the most serious reason behind SWI in developing regions [31]. Further research in the ND to assess the impact of CC versus extensive extraction as another responsible factor for salinization is needed.

2.4 GROUNDWATER IN THE NILE DELTA

2.4.1 Aquifer characteristics in the Nile Delta

The ND was extensively studied from geological, hydro-chemical and hydrological aspects. Many research studies have been implemented in the ND leading to identification of the characteristics of the aquifer.

The ND Quaternary aquifer is considered as a semi-confined aquifer [32]. It covers the whole ND. Its thickness varies from 200 m in the southern parts to 1000 m in the northern parts [33]. The depth to the groundwater table in this aquifer ranges between 1-2 m in the north, 3-4 m in the middle and 5m in the south [33]. [34] and [5] studied the characteristics of the NDA and declared that the top of the Quaternary aquifer is covered by a thin clay layer, which leads to the characterization of this main aquifer as a semi-confined aquifer. The thickness of the clay layer varies from 5-20 m in the south and the middle part of the ND, and reaches 50 m in the north [35]. The thickness and lithological differences of the clay layer have a great effect on the degree of hydraulic connection between the groundwater and surface water [36].

The main aquifer is formed by Quaternary deposits [32]. [5] attributed the variation of the hydraulic parameters and salinity of the aquifer to the fact that these deposits took place under different deltaic conditions. These deposits represent different aggradations and degradation phases that were usually accompanied with sea level changes [35]. The hydraulic connections among these deposits transformed the Quaternary aquifer to a large storage reservoir that is supplied directly by the Nile water through the extensive irrigation networks, especially in the southern part of the ND [36]. On the other hand, earlier investigations confirmed that there is no definite hydraulic connection between the Quaternary aquifer and the underlying Tertiary rocky deposits that act as an aquiclude [37].

Different hydraulic parameters of the main aquifer have been investigated by researchers. Table 2.1 summarizes the hydraulic parameters estimations of the NDA made by various authors. The high hydraulic conductivity values are attributed to the fact that the aquifer is composed mainly of sand and gravel [38]. Some parameters ranges are quite close across different studies, e.g. porosity. However, other parameter ranges are quite different, e.g. transmissivity. As indicated in Table 2.1, almost all of the studies gave an average value of hydraulic conductivity for the whole ND, which was subsequently used in further studies (including development of groundwater models). A regional area like the ND is characterized with spatially varying hydraulic conductivity for different locations and layers, which needs to be taken into account for more accurate representation of the study area. [5] published that vertical hydraulic conductivity of the clay layer is 0.0025 m /day while [39] documented it as 0.0011 m/day. With slightly higher values, [40] reported it at 0.0484 m/day and [41] at 0.0046 m/day. On the other hand, [29] used a vertical hydraulic conductivity about 0.67 m/day. Due to lack of data, subsequent studies used uniform value of vertical hydraulic conductivity all over the ND, not taking into consideration that the clay characteristics are spatially varying in the ND.

[42] stated that the average percolation to the Quaternary aquifer is about 0.8 mm/day. [43] published that the percolation rate ranges between 0.25 and 0.8 mm/day in the central and southern part of the delta, depending on the type of soil and irrigation and drainage practices. In the desert areas to the west, percolation rates which dominant range from 1.0 to 1.5 mm/day for furrow irrigation. They also found that the percolation rates in fields using drip and sprinkler irrigation ranged from 0.1 to 0.5 mm/day. The percolation rates ranged between 0.2 mm/day and 5 mm/day in the large reclamation projects in the eastern parts of the ND due to the subsurface drainage that prevailed [6]. Those percolation rates have been used widely in modeling studies.

Average rainfall in the ND is very small and ranges from 25 mm/year in the south and the middle part of the ND to 200 mm/year in the north [44]. From literature review, it can be concluded that the rainfall induced recharge is neglected in almost all groundwater modeling studies compared to the recharge from the returned irrigation flow.

Another significant influence to the recharge of the main aquifer comes from the water levels in the irrigation canals. These water levels are also a significant factor in groundwater modeling, because they influence the surface water-groundwater interaction [8]. The literature review shows that in most modeling studies these were represented with a constant average water level value along the canals. On the other hand, water levels of the canals vary from one month to another and throughout different sectors of the canals, which needs to be taken into account for more accurate representation of the interactions between the aquifer and the surface water in the ND.

The previous work that has been carried out has provided a better understanding of the aquifer. It has formed the basis for many researchers that have used the documented results as valuable input in groundwater modeling and simulation studies for different environmental problems that face the NDA. However, there is a gap in hydrological data series in the ND between different water sectors that works in the MWRI. Therefore, continuous monitoring of hydrological parameter could lead to more reliable research.

2.4.2 Groundwater salinization studies in the Nile Delta

Many researchers used chemical and isotopic analyses to detect the salinity of the groundwater aquifer as diagnostic tools for identifying the origin of the dissolved salts. SWI was the primary cause to explain the increase in salinity of groundwater especially in the northern parts. However, some other causes such as salinization coming from soil formations were also documented. [45] analysed the groundwater salinity and found that the range of groundwater salinity is between 227 ppm and 15,264 ppm. The lower salinity values are found in the southern parts of the ND region and near the canals of the Nile River due to soil salinity. His results agreed with the results of [5] that the northern zone is highly saline due to SWI. [31] analysed the historical records and concluded that the salinity of groundwater is changing with changing water levels of the canals. They mentioned that from 1957 till 1984, the groundwater salinity records showed that it was enhanced and the freshwater was dominating and overcoming SWI. They found that the groundwater heads were increasing during this period and they attributed that to the construction of High Aswan Dam because perennial freshwaters were delivered to the ND throughout the whole year. After 1984, the groundwater salinity started to increase due to extensive extraction and reduction in the flow of the Nile [31]. When the Nile water flow increased in 1990, the salinity of groundwater reduced again to its former levels [31]. However, in 2000, the salinity of groundwater increased again due to extensive extraction and new reclamation projects [31]. This interpretation of the historical data provided a clear general picture about the evolution of the Quaternary aquifer status in the ND.

The above mentioned researchers where among the pioneers from which a large number of subsequent researches branched. Chemical analyses by themselves are good tools to detect salinity in given conditions, but they are insufficient for forecasting future salinity conditions. Salinization analysis of the aquifer with all the hydrological dimensions is very complicated. Highly populated regions like the ND faced with a persisting issue of SWI require aquifer management based on prediction of future conditions that can be provided by groundwater modeling.

2.5 MODELING OF GROUNDWATER SALINITY

2.5.1 Groundwater salinity modeling studies using SEAWAT

A thorough overview of all aspects of groundwater SWI problems, including modeling approaches, is provided in the recent article of [46]. Therefore, we will not go in detailed overview of these modeling approaches, for which the readers are advised to access the mentioned reference. It is of importance, however, to mention that out of the two distinct approaches for modeling SWI, namely the sharp interface approach and the variable density approach, the applicability of the sharp interface approach for the integrated modeling of the NDA is quite limited. The reason for this is the fact that the transition zone between salt and fresh water in this aquifer (characterized with varying density) is quite large and needs to be captured by the intended model. [46] have tabulated the most widely used variable density codes. They documented the use of 2D/3D FEMFAT, FEFLOW, FEMWATER, HYDROGEOSPHERE, MARUN, MOCDENS3D, MODHMS, SUTRA, and SEAWAT by researchers. One of the most popular codes in recent years has been SEAWAT. Many references of usage of SEAWAT are listed in [46]. SEAWAT uses the concept of equivalent fresh water head for simulating density dependent flows, where the flow calculations are performed by the popular MODFLOW code and MT3DMS is used for the solute transport [47]. This code has shown very good results in SWI modeling studies in several different applications. Given its features and application potential, SEAWAT may be a good candidate code for developing the kind of integrated 3D model of the NDA that is argued for in this chapter. Some experiences with applications of SEAWAT are briefly presented as follows:

The original SEAWAT code was written by [47] referred to as version 1. It was applied to simulate groundwater flow and SWI in coastal environments. It was modified by [38]. [49] presented the formal documentation for version 2 of SEAWAT code. [50] implemented SWIFT2D coupled with SEAWAT to simulate the hydrological processes in coastal wetlands. They concluded that the integrated code gave very good results and could be widely used in SWI problems. Afterward, [51] conducted a study to evaluate the relation between water-level fluctuations and SWI in Broward County, Florida, using SEAWAT. The model was used to simulate movement of the saltwater interface resulting from changes in precipitation, extraction, sea-level movement, and upstream canal stage. The results indicated that the canal control structure and sea level have major effects on groundwater flow. They concluded that SEAWAT code provides very reliable results. [52] used SEAWAT code to analyze freshwater and saltwater flow. They found that the subsurface geology greatly affects the position and movement of the underlying freshwater/saltwater interface. Moreover, the authors concluded that pumping from large-capacity municipal-supply wells increases the potential of impacts on surface-water resources that are affected by pumping and wastewater disposal locations.

These studies indicate that SEAWAT has been successfully used for model-based analysis of a wide range of SWI problems that have similar characteristics to those in the NDA. Like with other variable density codes the main problems that researchers could face when using SEAWAT are in determining the right trade-off between required complexity that is needed for interpreting the predicted salinity distribution and long running times, and the efforts needed for model calibration. Nevertheless, such modeling codes have allowed possibilities for simulating 3D variable-density groundwater flow and predicting the magnitude and direction of SWI under changed future conditions.

2.5.2 Groundwater salinity modeling studies in Egypt

Various numerical techniques were used to assess and simulate the SWI in the ND. Earlier studies were mainly focused on determining the freshwater thickness of the NDA using (semi- analytical) models based on the sharp interface modeling approach. Examples of such studies can be found in [34, 53, 54, 31]. Most of these studies were rather theoretical in nature as there were not enough records of salinity of the aquifer. As we have mentioned earlier, in case of the ND the transition zone is relatively large and characterized by the dynamic relation between fresh and seawater. Consequently, the variable density numerical models are better suited for simulating the interactions of the freshwater and seawater in the aquifer. In recent years, such models have been developed either as 2D vertical models for selected cross sections of the ND, or 2D horizontal models for parts of the NDA.

In Egypt, extensive unplanned extraction causes the deterioration of the Quaternary aquifer, especially in the northern coast [31]. Historical records show a continuous increase in the extraction rates over the last 30 years (during the period of 1980-2010). In 1980, the Research Institute of Groundwater in Egypt (RIGW) launched a primary study to estimate the safe yield of the NDA [55]. 2D finite difference models were applied to determine the effect of extraction on the water levels and the safe yield of the NDA. However, these models did not take into account the SWI phenomena. The research declared that the total annual extraction rate in 1980 was estimated at about $1.6x10^9$ m^3/year. In addition, the net recharge rate to the Quaternary aquifer was estimated to $2.6x10^9$ m^3/year. The results from chemical analyses of the groundwater did not show increase in its salinity, in spite of the reduction in the amount of annual outflow to the sea and the increase of extraction rates, compared to rates of extraction in 1960. Consequently, the study concluded that both salt and freshwater status was in dynamic equilibrium. The study recommended that the annual extraction rates should increase by $0.5x10^9$ m^3/year. They attributed this to the need to lower the groundwater head in order to prevent water logging and soil salinization. [53] used a 2D finite element model called AQUIFEM1 based on movable sharp interface depending on extraction. The model results estimated an optimal annual groundwater extraction that should not exceed $4.8x10^9$ m^3/year. Official reports from RIGW confirmed significant increase in patterns of extraction, which reached around $2.6x10^9$ m^3/year in 1991. The numbers of wells have doubled from 1958 to 1991 [44]. In 1999, a project entitled "Water Resources Management under Drought Conditions" studied the Nile Valley and the NDA system using the TRIWACO model code, a finite element variable density numerical model. They found that, there is an alarming danger that urgently needs a comprehensive management plan for drought mitigation based on limiting extraction rates all over Egypt. They noted that the annual extraction reached around $3.02x10^9$ m^3/year in the ND. In 2003, the total annual extraction reached $3.5x10^9$ m^3/year$^-$ [8]. In 2010, it reached about 4.9×10^9 m^3/year [8]. Following the trend of the increase of extraction in the ND, it can be noticed that it increases linearly by about $0.1x10^9$ m^3/year, except from the period of 2003 till 2010 where the extraction increases dramatically by rate of $0.2x\ 10^9$ m^3/year.

A number of modeling studies focused on analyzing the impact of increased groundwater extraction on the salinization of the NDA [56] used the SUTRA model code to simulate the behavior of the transition zone of the ND under different groundwater extraction intensities. He declared that the northern part of the middle ND is more salinized than the southern part. The model tested the impact of pumping freshwater and brackish water simultaneously which is known as the scavenger well scheme. He concluded that a unique saline well could be used in order to control a number of four or more fresh water pumping wells at a certain distance (circle of influence) to maintain the transition zone at its equilibrium position. [57] studied SWI in the NDA under the effect of fresh water storage in the northern lakes of Manzala and Burullus. The authors simulated the system using SUTRA model and Lake Model. They confirmed that there is SWI in the northern part where the fresh water of the lakes minimizes the intrusion around their zone of influence.

Among the scientist that adopted the variable density approach to study SWI were [58, 59, 60, 61, 62]. They outlined the freshwater-seawater interface in the horizontal and vertical cross sections. [63] studied the impact of CC on the Quaternary aquifer of the ND and compared it with the coastal aquifer in India. They modeled both aquifers and assumed three most likely scenarios for SLR. They found that the NDA is more vulnerable than the coastal aquifer in India to SLR. They attributed that, to the elevation of the coastal land in the ND which is very low. In addition, the coastal area is subjected to land subsidence that could cause more SWI. Most recently, [29] discussed the concept of equivalent freshwater head in successive horizontal simulations of SWI in the ND. The authors used FEFLOW, a 3D finite element variable density model. However, due to the unavailability of data, the simulations were performed as 2D sequences (vertical layers). Their results clearly demonstrate that the location of the transition zone moves towards landside as moving down with depth. They found that in the middle of the ND the SWI reached inland to a distance of 40 km. It is followed by 30 km transition zone. The width of the transition zone reached its minimum value (about 6 km) in Damanhour city. [64] developed a full 3D model, using 28 vertical layers, but most of the assumptions about hydrological stresses remained the same as in their previous work of [18]. However, more research is needed with reliable hydrological data oriented towards development of a fully 3D variable density model of the NDA that can serve as a predictive tool for analyzing future mitigation and adaptation measures

2.6 MITIGATION AND ADAPTATION MEASURES

In case of the ND, existing studies were predominantly focused on adaptation measures. Very few existing studies have discussed mitigation measures related to groundwater salinization. Mitigation measures were more studied in relation to the erosion of the coastal strip of the ND, which is another problem that can be increased in the future due to SLR and more severe weather events. Table 2.2 summarizes a number of adaptation and mitigation measures proposed by different researchers and their advantages and disadvantages.

Most of the work that has been carried out in the above proposed adaptive measures is directed towards a specific location in the ND. The disadvantage of this is that the proposed adaptation plan could negatively influence another region of the ND. Unfortunately, most of the proposed adaptation and/or mitigation measures in the ND stop at the phase of recommendation. A comprehensive strategy for adaptation measures that is proposed as a result of model-based analysis and evaluation is missing. Also, the effect of integrating two or three adaptation methods together has not been studied. Model-based analysis of such combinations may indicate a possible way forward. In addition, strong institutional capabilities to implement some of the proposed alternatives could be a large constraint in Egypt, as the case in many developing countries. The need for embedding proposed adaptation and mitigation measures in a broader groundwater management strategy will be further addressed in the following section.

2.7 Discussion

As we have seen, CC and its impact on the ND was the subject of comprehensive studies for the past 30 years. Most of the research studies were focused on determining the impact of CC on precipitation and temperature patterns affecting the Nile basin flows that are critical for the ND. However, the results from these studies are far from conclusive and further research is needed. Studies of SLR due to CC have mostly focused on quantifying impacts on the ND coast and surface water. Integrated groundwater model of the whole NDA that includes freshwater-saltwater interactions could serve as a one of the possible tools for the quantification and characterization of these impacts.

Increased and largely uncontrolled groundwater extractions are potentially more serious threat to the salinization of the NDA. Historical trends demonstrate continuous increase of groundwater extractions over the last three decades. Most modeling studies documented in literature simulated the NDA for studying the deterioration and salinization of the aquifer due to this already recognized threat. At the same time, however, the majority of reported modeling studies were of local nature, implemented in specific regions to analyse the problems of a particular zone and interpret the results in terms of impacts caused by local cause's extractions. However, all ND parts are connected and should be integrated together in order to identify their relations and influences. A comprehensive substantial regional analysis of the whole ND is limited in the literature.

Past modeling work was mainly carried out using 2D vertical (cross sectional) or 2D horizontal models. An important restriction of 2D vertical models is that the representative cross sections should be selected carefully and that results are not transferable to other areas of the ND. The 2D horizontal models provide spatial results in horizontal dimensions, but they give less accurate location and shape of the transition zone between fresh and saltwater in vertical dimension. This situation indicates that future research should focus on development of fully 3D models, but due to their complexity, data needs and long computational time, such models are rarely developed for SWI problems, and no such model has been developed for the whole NDA. Yet, such models are clearly needed. There are a number of researchers who have successfully used 3D modeling in a large computational area worldwide and this model proved to give promising results, e.g. [65] who used MOCDENS 3D to simulate the coastal lowlands of the Netherlands. They calculated the possible impacts of future SLR, land subsidence, changes in recharge, autonomous salinization, and the effects of two mitigation countermeasures with a 3D numerical model for variable density groundwater flow and coupled solute transport. In addition, [66] successfully used a 3D simulation of SWI in a case study in the Goksu Deltaic Plain. The 3D modeling helped them in better understanding the behavior of SWI in the coastal aquifer.

Consequently, the 3D modeling can be used for hypothesis testing and for better understanding of the overall system behavior of the NDA in 3D. Moreover, these models can be used for assessment of the degree of salinization of the whole NDA. Finally, once fully developed, such models can become central components of future planning platforms and decision support systems for evaluation of different adaptation and mitigation measures.

A significant problem that prevents scientists from advancing research in the ND is the lack of data of sufficient quality. Wells that monitor SWI (especially deep wells) are lacking [29]. Drilling a number of deep wells that cover the NDA would provide additional information on the salinity of the deep zone of the aquifer. Such salinity data are needed for calibration and validation of models and without these data the accuracy of modeling results remains doubtful.

Data gathering campaigns in Egypt are usually temporary in nature, depending on available funding for particular projects. Furthermore, as in many other countries, existing data are available from different agencies and other organizations and their collection requires considerable effort. It can be argued, however, that the development of a 3D integrated model for the NDA needs to start using all presently available data. Even if the quality of such model would be somewhat impaired because of lack of data, the model itself can be of assistance for designing and implementation of the needed monitoring system, e.g. by identifying critical, vulnerable areas which require denser monitoring network.

Identifying hotspots for some parts of the ND has been addressed by few researchers, either by using modeling results [27] or by GIS analyses of existing data [67]. Although there is extensive extraction in the ND, little is known in combined influences of SLR and development-related groundwater extractions.

Regarding adaptation measures, the analysis of previous studies shows that at regional scale is not addressed. Studies are rarely based on a numerical assessment covering both natural and associated social and economic changes. In fact, the adaptation measures need to be analyzed within an integrated regional plan. Various measures proposed (e.g. hydraulic barrier, physical barrier, air or fresh water injection, etc.) should then be extensively studied with the view of economic perspective and applicability. For example, in case of hydraulic injection, the availability and type of injected water, or in case of extraction of brackish water, the method of disposal of this water without harming the ecosystem, should be known. A comparative performance study could then be carried out taking into account the time that each method would take for completing the mitigation and remediation. Multi-criteria analyses could be carried out by taking into account the complete economic and environmental feasibility objectives for a whole set of measures and these results could then be presented to decision makers. This type of research would link numerical modeling results with socioeconomic constraints together with ecosystem interaction as this has been so far highly neglected. These objectives could be achieved by integrating all required components (data, numerical models, multi criteria analysis tools) in a comprehensive Decision Support System (DSS) that can be used by the relevant authorities and stakeholders.

The envisaged groundwater management plan within which the effectiveness and feasibility of the proposed measures is to be assessed should include additional measures for controlling the excessive extractions. Although development of legal regulations and associated strategies (e.g. to restrict development in areas vulnerable to salinization) should provide the broad framework for this approach, this aspect has not been addressed in existing research.

2.8 AVENUES FOR FUTURE RESEARCH

We conclude this chapter with several ideas about the possible future research directions that may provide useful inputs for sustainable management of groundwater resources from the NDA and prevent further salinization problems. From the previous section it becomes apparent that the key research activities in future would be reasonable to aim at developing a regional 3D variable density numerical model of the NDA. It should be clear that developing a 3D model is only one of the tools that could be used to understand the behavior of the NDA. The multiple benefits of developing such a model have been presented in the previous sections. The popular SEAWAT model code, based on MODFLOW and MT3DMS is a good candidate for setting up such a model.

The proposed model should cover the whole ND. Given the large area of the ND, the envisaged model would still have rather coarse discretization in horizontal direction (1-2 km per grid cell). According to the hydrological settings, the aquifer contains two main layers, the Holocene and the Pleistocene as mentioned before. However, sufficiently fine discretization in vertical direction is needed for capturing the transition zone between fresh and seawater (10-30 vertical layers). Such decisions will need to be made by considering the length of computational time, which is an issue for most SWI models. It should be noted that, the coarse discretization will not allow for accurate prediction of salinity at specific sites, however, it will give a regional perspective for the freshwater and saline water interface. In addition to 3D modeling to the ND, running simpler models in different governorates in the ND is also advisable. This will give improved insight of the salinity accurately at local areas.

First step in this research would be collection and synthesis of all existing data. Having a complete set of data series is especially problematic in Egypt, but it is important that enough reliable and updated data sets on different physical, hydrological and hydro-geological variables and parameters are gathered. This will enable the development of the conceptual and the numerical model of the ND. This model could also be used for analysis of future conditions through scenario simulations. In addition to producing a useful predictive tool, the model development process would contribute to improving our understanding of the hydro-(geo)logical processes in the NDA and especially of the processes related to SWI.

Given that the main drivers for further deteriorating impacts on the aquifer are identified to be CC (SLR) and increased development-driven groundwater extraction, further analysis with the developed model should be carried out on quantifying changes in salinity conditions in the aquifer as induced by these two external drivers. Existing CC scenarios could be used to formulate possible future SLR and hydrological conditions, while development plans within Egypt could offer information for estimating future levels and spatial distribution of groundwater extractions.

Special emphasis should be put on extreme conditions/combinations of SLR and groundwater extractions. The results of the model-based analysis could serve as primary targets for introduction of possible future mitigation and adaptation strategies.

The possible effects of mitigation/adaptation measures on the basis of the model-generated results could then be quantified and different measures evaluated. Strategies for groundwater conditions could be assessed with and without combinations of adaptation measures. The most appropriate adaptation and/or mitigation scenarios could then be recommended to be considered for implementation.

These proposed lines of research will contribute to a comprehensive framework for development of long-term planning for sustainable management of groundwater resources in the NDA. Once the model is available and its usefulness is confirmed through the applications described above, further steps could be made towards development of a DSS for groundwater planning and management in the NDA, as discussed in section 6. This will encompass many more than, only physical aspects of groundwater management, for instance, management options for the conjunctive use of water resources, socio-economic assessment of alternatives etc.

The model proposed for the NDA could also serve as an example for similar areas around world. New insights provided by this research may lead to application of the investigated methodologies in other comparable deltas, since the problems that the ND faces nowadays will very likely be future problems encountered in other deltas elsewhere in the world [68, 69].

Table 2.1. Reported hydraulic parameters of the Quaternary aquifer in the Nile Delta

References	Transmissivity (m^2 day)	Hydraulic Conductivity (m/day)	Porosity (%)
[70]	71,800	86	26
[5]	72,000	112	40
[71]	20,000–103,000	55–103	------
[72]	------	72–108	21-30
[73]	------	119	30
[74]	2,500–25,900	50	25
[6]	10,350–59,800	150	25–30
[44]	15,000–75,000	35–100	25–40
[75]	------	25–40	25
[76]	------	23–65	25–40
[77]			
southern (ND)	5,000-25,000	50-100	25-30
northern (ND)	------	Less than 50	More than 30
[29]	2,000-15,000	36-240	25-40

Table 2.2. Advantages and disadvantages of different adaptation and mitigation measures for groundwater salinization in deltas worldwide

Measure	Advantage	Disadvantage	Conclusion
a. Adaptation			
1.Rice cultivation [78, 30]	Soil salinization patterns decrease considerably	Needs a large amount of water which is already a scarce resource.	Not recommended as it is often uneconomic and not environemtally sustainably
2. Permitting 10 to 20% of the freshwater of irrigation to leach the soil. [79]	No salt accumulation, salt export will match salt import and will eventually prevent salt infiltration to groundwater.	This could be risky because it might cause salt returning to the root zone again.	Not recommended in most cases
3. Cultivating salt tolerant crops. [80]	Tolerant crops can withstand salt concentration in the north	Very limited types of plants.	The salinization will increase continuously and this is only a temporary solution
4. Creating wetlands in salinized areas. [1]	Egypt has four lakes in the northern coast of the ND which could be considered as natural adaptation.	Only applicable in low lying parts of the ND cannot be used in agriculture	Recommended to lower salinization levels in some parts of Egypt
5. Extraction of saline groundwater. [81]	Getting rid of saline water.	Disposal of extracted saline water could cause another environmental problem.	Not recommended in shallow coastal aquifers
6. Increasing land reclamation. [81]	Increase freshwater recharge	Need of land and fresh water	It is recommended, if resource are available
b. Mitigation measures			
1.Artificial recharge [82, 83, 84]	Increase freshwater outflow to the aquifer. The degree of efficiency of this method depends on pumping / injection rates, depth of the wells, the coastal aquifer properties and the location of the	Needs a large amount of water, which is already a scarce resource.	It is recommended in case of water abundance as it is a highly effective method.

Measure	Advantage	Disadvantage	Conclusion
	wells.		
2.Physical barriers [85, 81]	This method stabilizes the coast and decreases SWI. The height of the barrier has a very significant role in the degree of flushing rates.	It is one of the most expensive methods either using sheet piles or clay trenches. Nevertheless, it is only applicable in shallow aquifer because of its large cost	Economic feasibility is the cornerstone
3.Air injection [86, 46]	This method minimizes the aquifer permeability.	In experimental stage and not fully developed.	Further experiments on bigger scale are needed

REFERENCES

1 Oppenheimer, M., B.C. Glavovic, J. Hinkel, R. van de Wal, A.K. Magnan, A. Abd-Elgawad, R. Cai, M. Cifuentes-Jara, R.M. DeConto, T. Ghosh, J. Hay, F. Isla, B. Marzeion, B. Meyssignac, and Z. Sebesvari, 2019: Sea level rise and implications for low-lying islands, coasts and communities. In: IPCC special report on the ocean and cryosphere in a changing climate, 2019.

2 El Raey, M., Fouda, Y., and Nasr, S., GIS assessment of the vulnerability of the Rosetta area, Egypt to impacts of sea rise, *Environ. Monitor. Assess.J.,* 47, 59–77, 1997.

3 EGSA., Egyptian General Survey and Mining: Topographical Map cover ND, scale 1: 2 000 000, 1997.

4 SADS2030., Sustainable agricultural development strategy, Egypt, Ministry of Agriculture and Land Reclamation,1st eddition,197,2009.

5 Farid, M. S. M., Nile Delta groundwater study, M.Sc. thesis, Cairo Univ, Egypt, 1980.

6 Leaven, M. T., Hydrogeological study of the ND and adjacent desert areas, Egypt, with emphasis on hydrochemistry and isotope hydrology, thesis, Free Univ, Amsterdam, also published by RIGW/IWACO as Technical note TN 77.01300-91-01, 1991.

7 Dawoud, A. M.: Design of national groundwater quality monitoring network in Egypt, *Environ. Monitor. Assess. J.* 96, 99–118, 2004.

8 RIGW: Annual report for year 2010, Research Inst. for Groundwater, Kanater El-Khairia, Egypt, 2010.

9 El Sayed, M. Kh. Implications of climate change for coastal areas along the ND. *The Environ. Profess. J.,* 13, 59-65, 1991.

10 Nossar, M. Climatic changes and their impacts on groundwater occurrence in the eastern part of ND, Egypt, PhD. thesis, Zagazig Univ, Egypt, 2011.

11 Di Baldassare, G. D., Elshamy, M., Griensven, A. V., Soliman, E., Kigobe, M., Ndomba, P., Mutemi, J., Mutua, F., Moges, S., Xuan, Y., Solomatine, D., and Uhlenbrook, S.: Future hydrology and climate in the River Nile basin: a review, *Hydrol. Sci. J.* 56, 199–211, 2010.

12 Strzepek, K. and Yates, D. N., Economic and social adaptation to CC impacts on water resources: a case study of Egypt, *Water Res. Devel. J.* 12, 229–244, 1996.

13 Strzepek, K. M., Yates, D. Yohe, G., and Tol, R. J., Constructing not implausible climate and economic scenarios for Egypt, *Integrat. Assess. J.* 2, 139–157, 2001.

14 Sestini, G.: The Implications of CCs for the ND, report WG, 2/14, Nairobi, Kenya: UNEP/OCA, 1989.

15 Frihy, O. E, Deabes, E.A.,Shereet, S.M.,Abdalla, F.A., ND coast,Egypt-Update and future projection of relative SLR, *Environ. Earth Sci. J.* 61, 1866-6299, 2010.

16 Syvitski, J.P.M., Kettner, A.J. Overeem, I., Hutton, E.W.H., Hannon, M.T., Brakenridge, G.R., Day, J., Všršsmarty, C., Saito, Y., Giosan, L., and Nicholls, R.J., Sinking deltas due to human activities, *Nat. Geosci. J.* 2, 681–686, 2009.

17 Emery, K. O., Aubrey, D. G., and Goldsmith, V., Coastal neotectonics of the Mediterranean from tide-gauge records, *Mar. Geol. J.* 81, 41–52, 1988.

18 Stanley, D. J. Recent subsidence and northeast tilting of the ND, Egypt, *Mar. Geol.J.* 94, 147–154, 1990.

19 Frihy, O. E., Beach response to SLR along the delta coast of Egypt, Int. *Union of Geodesy and Geophysics and the American Geophysical Union J.* 11, 81–85, 1992.

20 Eid, H., El-Marsafawy, S., and Ouda, S., Assessing the economic impacts of CC on agriculture in Egypt, policy research working paper 4293, The World Bank Development Research Group, 2007.

21 Alam El Din, K. A. and Abdel Rahman, S. M., Is the rate of SLR accelerating along the Egyptian coasts, CC impact in Egypt, adaptation and mitigation measures conference, Egyptian Research Center, Alexandria, Egypt, 2009.

22 CRI/UNESCO/UNDP., Coastal protection studies, final technical report, 155, Delft, 1978.

23 Delft Hydraulics and Resource Analysis, Vulnerability assessment to accelerated SLR, case study Egypt. Final report prepared for commission of European communities in cooperation with the Coastal Research Institute, Alex., Egypt, 1992.

24 El Fishawi, N. M., Recent sea level changes and their implications along the ND coast, sea level change and their consequences for hydrology and water management noordwijkerhout, the Netherlands, UNESCO, IHP-IV Project H-2-2.PP.IV.3-11, 1993.

25 Stanley, D. J. and Warne, A. G., ND, Recent geological evolution and human impact, *Sci.J.* 260, 628–634, 1993.

26 El Raey, M., Nasr, S., Frihy, O., Desouki, S., and Dewidar, Kh., Potential impacts of accelerated sea-level rise on Alexandria governorate, Egypt, *Coast. Res. J.,* 14, 190–204, 1995.

27 El Raey, M., Frihy, O., Nasr, S. M., and Dewidar, K. H., Vulnerability assessment of SLR over Port said governorate, Egypt, Kluwer Academic Publishers, *Environ. Monitor. Assess. J.* 56, 113–128, 1999.

28 Ferguson, G. and Gleeson, T.: Vulnerability of coastal aquifers to groundwater use and CC, J. Nature CC, 2, 342–345, 2012.

29 Sherif, M. M., Sefelnasr, A., and Javad, A., Incorporating the concept of equivalent freshwater head in successive horizontal simulations of seawater intrusion in the NDA, Egypt, *Hydrol. J.* 464-465,186-198, 2012.

30 Kotb, T. H. S., Watanaba, T., Ogino, Y., and Nakagiri, T., Soil salinization in the ND and related policy issues in Egypt, *Agr. Water Manage. J.* 43, 239–261, 2000.

31 Sakr, S. A., Attia, F. A., and Millette, J. A., Vulnerability of the NDA of Egypt to seawater intrusion, International conference on water resources of arid and semi-arid regions of Africa, Issues and challenges, Gaborone, Botswana, 2004.

32 Ball, J., Contribution to the Geography of Egypt, Surv., Cairo, Egypt. Book, 23–84, 1939.

33 Dahab, K., Hydrogeological evaluation of the ND after High Dam construction, Ph.D. thesis, Fac. of Sci., Menoufia Univ, Egypt, 1993.

34 Wilson, J., Townley, L. R., and Sa Da Costa, A., Mathematical development and verification of a finite element aquifer flow model AQUIFEM-1, technology adaptation program, report No. 79–2, M.I.T., Cambridge, Massachusetts, 1979.

35 Diab, M. S., Dahab, K., and El Fakharany, M., Impacts of the paleohydrological conditions on the groundwater quality in the northern part of ND, The geological society of Egypt, *Geol. J.* 4112B, 779–795, 1997.

36 Abdel Maged, M. S. H.: Water logging phenomena in the north of the delta region, MSc. thesis, Fac. of Sci., Cairo Univ, Egypt, 1994.

37 Saleh, M. F., Some Hydrological and Hydrochemical Studies on the ND, MSc. thesis, Fac. of Sci., Ain Shams Univ, Egypt, 1980.

38 Marotz, G., Tecnische grunlageneiner wasserspeicherung im naturlichen untergrund – schriftenreihe des 5 KWK, 18, 228, 95 Abb., 14 Tab., Anlage: Hamburg (Wsser U. Boden), 1968.

39 Wolf, P.: The problem of drainage and its solution in the Nile Valley and ND, *Natural Res. Develop. J.* 25, 62–73, 1987.

40 RIGW/IWACO, hydrological inventory and groundwater development plan western ND region, TN77. 01300-9-02 Research Inst. for Groundwater, Kanater El-Khairia, Egypt, 1990.

41 Arlt, H. D., A hydrogeological study of the NDA with emphasis on SWI in the northern Area, Mitteilung/Institut fur Wasserbau und Wasserwirtschaft, Technische Universit at Berlin, Nr. 130, OCLC No. 636899992, 291–302, 1995.

42 DRI, Drainage Research Institute, Cairo, Egypt: Land Drainage in Egypt edited by: Amer, M.H. and de Ridder, N. A., Drainage Research Institute, Cairo, Egypt, 1989.

43 Warner, J. W., Gates, T. G., Attia, F. A., and Mankarious, W. F., Vertical leakage in Egypt's Nile Valley: estimation and implications, *Irrig. Drain Eng.-ASCE J.* 117, 515–533, 1991.

44 RIGW: Research Institute for Groundwater, Hydrogeo. map of ND, Scale 1: 500,000, 1st Edn., ND, 1992a.

45 Atta, A. S., Studies on the groundwater properties of the ND, Egypt, M.Sc. thesis, Fac. of Sci., Cairo Univ, 311–325, 1979.

46 Werner, A. D., Bakker, M., Post, V.E.A., Vandenbohede, A., Lu, C., Ataie-Ashtiani, B., Simmons, C., T. Barry, and D.A., Seawater intrusion processes, investigation and management: Recent advances and future challenges, *Adv. Water Resour. J.* 51, 3-26, 2013.

47 Guo, W. and Bennett, G. D., Simulation of saline/fresh water flows using MODFLOW, edited by: Poeter, E. P., Zheng, C., and Hill, M. C., Proceedings of the MODFLOW '98 Conference, Golden, Colo., 1, 267–274, 1998.

48 Langevin, C. D. and Guo, W., Improvements to SEAWAT, a variable-density modeling code [abs.], EOST. 80, no. 46, F-373, 1999.

49 Guo, W. and Langevin, C. D., User's Guide to SEAWAT: A computer program for simulation of 3D variable-density groundwater flow, techniques of water-resources investigations book 6, Ch.7, 77, 2002.

50 Langevin, C. D., Oude Essink, G. H. P., Panday, S., Bakker, M., Prommer, H., Swain, E. D., Jones, W., Beach, M., and Barcelo, M., Ch.3, MODFLOW-based tools for simulation of variable-density groundwater flow: in coastal aquifer management: monitoring, modeling, and case studies, edited by: Cheng, A. and Ouazar, D., Lewis Publishers, 49–76, 2004.

51 Dausman, A. M. and Langevin, C. D., Movement of the saltwater interface in the surficial aquifer aystem in response to hydrologic stresses and water-management practices, Broward County, Florida, US Geological Survey Scientific Investigations Report, 52–56, 2004.

52 Masterson, J. P. and Garabedian, S. P., Effects of sea-level rise on groundwater flow in a coastal aquifer system, *Groundwater J.* 45, 209–217, 2007.

53 Farid, M. S.: Management of groundwater system in the ND, Ph.D. thesis, Fac. of Eng., Cairo Univ, Egypt, 1985.

54 Amer, A. and Farid, M. S., Sea water intrusion phenomenon in the NDA, Proceed. of the int. workshop on Management of the ND ground Aquifer, CU/MIT, Cairo, 1981.

55 RIGW: Project of safe yield study for groundwater aquifers in the ND and upper Egypt, part 1, Ministry of Irrig., Academy of Sci. Res. & Tech., and Org. of Atomic Energy, Egypt, 1980 (in Arabic).

56 Gaame, O. M.: The behavior of the transition zone in the NDA under different pumping schemes, Ph.D. thesis, Fac. of Eng., Cairo Univ, Egypt, 2000.

57 El Didy, S. M. and Darwish, M. M., Studying the effect of desalination of Manzala and Burullus Lakes on salt water intrusion in the ND, *Water Sci. J.* National Water Research Center, 2001.

58 Sherif, M. M., Singh, V. P., and Amer, A. M., A 2D finite element model for dispersion (2D-FED) in coastal aquifer, *Hydrol. J.* 103, 11–36, 1988.

59 Sherif, M. M., Singh, V. P., and Amer, A. M., A note on SWI in coastal aquifers, *Water Resour. Manage. J.* 4, 113–123, 1990.

60 Darwish, M. M., Effect of probable hydrological changes on the NDA system, Ph.D. thesis, Cairo Univ, 1994.

61 Sherif, M. M. and Singh, V. P., Groundwater development and sustainability in the NDA, final report submitted to Binational Fulbright Commission, Egypt, 1997.

62 Sherif, M. M., The NDA in Egypt, Ch. 17 in seawater intrusion in coastal aquifers, concepts methods and practices, edited by: Bear, J., Cheng, A., Sorek, S., Ouazar, D., and Herrera, A.: Theory and application of transport in porous media, Kluwer academic publishers, the Netherlands, 14, 559–590, 1999a.

63 Sherif, M. M. and Singh, V. P., Effect of CC on seawater intrusion in coastal aquifers, *Earth sci. J.* Hydrological processes, 13, 1277–1287, 1999.

64 Sefelnasr, A., Sherif, M.M., Impacts of seawater rise on seawater intrusion in the NDA, Egypt. *Groundwater J.* 52, 2, 264-276, doi 10.1111/gwat.12058, 2014.

65 Oude Essink, G. H. P., van Baaren, E. S., and de Louw, P. G. B., Effects of CC on coastal groundwater systems: a modeling study in the Netherlands, *Water. Resour. J.* 46, W00F04, doi: 10.1029/2009WR008719, 2010.

66 Cobaner, M., Yurtal, R., Dogan, A., & Motz, L. H., 3D simulation of seawater intrusion in coastal aquifers: A case study in the Goksu Deltaic Plain. *Hydrol. J.* 464-465, 262–280, doi: 10.1016, 2012.

67 Morgen, S. and Shehata, M., Groundwater vulnerability and risk mapping of the Quaternary aquifer system in the northeastern part of the ND, Egypt, *Int. Res. Geol. & Min J.* (IRJGM) (2276-6618), 2, 161–173, http: //www.interesjournals.org/IRJGM, 2012.

68 Bucx, T., 5 Marchand, M., Makaske, A., and van de Guchte, C., Comparative assessment of the vulnerability and resilience of 10 deltas, synthesis report, Delta Alliance report no. 1, Delta Alliance International, Delft-Wageningen, the Netherlands, 2010.

69 De Vries, I., de Vries, A., Veraart, J. A., Oude Essink, G. H. P., Zwolsman, G. J., Creusen, R., and Buijtenhek, H. S., Policy options for sustainable fresh water supply in saline Deltas areas, Int. Conference, Deltas in time of CC, Rotterdam, Netherlands, 2010.

70 Shata, A. A. and El Fayoumy, L. F., Remarks on the hydrology of the ND, UAR, proceed. of the Bucharest Symposium, IASHIAIHS-UNESCO, 1970.

71 UNDP: Competitive use of water by major field crops in Egypt, Water Master Plan, Cairo, Egypt, soil and Water Research Institute, Ministry of Irrigation/UNDP/IBRD, 28, 1981.

72 Mabrook, B., Swailem, F., El Sheikh, R., and El Dairy, F., Shallow aquifer parameters and its influence on groundwater flow, ND Egypt, Australian Water Resour., Council Conf., Series 8, 187–197, 1983.

73 Zaghloul, M. G., Flow distribution through groundwater aquifer of the ND, M.Sc. Thesis, Fac. of Eng., Alex. Univ, Egypt, 1985.

74 Shahin, M., Hydrology of the Nile basin development in *Water Sci. J.* 21, Amsterdam, Netherlands, Elsevier Sci. Publishers, B.V., 575, 1985.

75 Bahr, B., NDA with emphasis on SWI in the northern area, MSc. thesis, Technical Univ of Berlin, Inst. for Applied Geosci., Berlin, Germany, 1995.

76 Sollouma, M. and Gomaa, M. A. Groundwater quality in the Miocene aquifers East and West of the ND and the north west Desert, Egypt, *Scientific J. of Fac. of Sci.* Ain Shams Univ., Cairo, Egypt, 35, 47–72, 1997.

77 RIGW/IWACO: Environmental management of groundwater resources (EMGR), Final technical report TN/70.0067/WQM/97/20, Research Inst. for Groundwater, Kanater El-Khairia, Egypt, 1999.

78 El Gunidy, S., Risseeuw, I. A., and Nijland, H. J. Research on water management of rice fields in the ND, Egypt, Int. inst. for land reclamation and improvement/ILRI Wageningen, Netherlands, Vol. 41, 72, 1987.

79 Abrol, I. P., Yadav, J. S. P., and Massoud, F. Salt affected soils and their management, Food and Agri. Org. of the United Nations (FAO), *Soils Bull.,* 39, 57–60, 1988.

80 FAO: Water quality for agriculture irrigation and drainage, paper no. 29, rev. 1, Rome, 1985.

81 Oude Essink, G. H. P., Improving fresh groundwater supply problems and solutions, *Ocean Coast. Manage. J.* 44, 429–449, 2001.

82 Bray, B. S. and Yeh, W.W. G., Improving seawater barrier operation with simulation optimization in southern California, *Water Resour. J.* Pl. Manage., 134, 171–180, 2008.

83 Luyun, R., Momii, K., and Nakagawa, K., Effects of recharge wells and flow barriers on seawater intrusion, *Ground Water J.* 49, 239–249, 2011.

84 Carrera, J., Hidalgo, J. J., Slooten, L. J., and Vazquez-Sune, E.: Computational and conceptual issues in the calibration of seawater intrusion models, *Hydrogeol. J.* 18, 131–145, 2010.

85 Fanos, A. M., Khafagy, A. A., and Dean, R., Protective works on the ND coast, *Coast. Res. J.* 11, 516–528, 1995.

86 Dror, I., Berkowitz, B., and Gorelick, S. M., Effects of air injection on flow through porous media: observations and analyses of laboratory-scale processes, *Water Resour. J.* 40, W09203, doi: 10.1029/2003WR002960, 2004

3 ASSESSING THE FRESH-SALINE GROUNDWATER DISTRIBUTION IN THE ND AQUIFER USING A 3D VARIABLE-DENSITY GROUNDWATER FLOW MODEL

The content of this chapter is a full reproduction of the published article:

Mabrouk, M.; Jonoski, A.; Oude Essink, G.H.P.; Uhlenbrook, S. Assessing the Fresh–Saline Groundwater Distribution in the Nile Delta Aquifer Using a 3D Variable-Density Groundwater Flow Model. *Water J.* 11, 1946, 2019.

3.1 Abstract

The Nile Delta Aquifer (NDA) is threatened by saltwater intrusion (SWI). This chapter demonstrates an approach for identifying critical salinity concentration zones using a three-dimensional (3D) variable-density groundwater flow model in the NDA. An innovative procedure is presented for the delineation of salinity concentration in 2010 by testing different simulation periods.

The results confirm the presence of saline groundwater caused by SWI in the north of the NDA. In addition, certain regions in the east and southwest of the NDA show increased salinity concentration levels possibly due to excessive groundwater extraction and dissolution of marine fractured limestone and shale that form the bedrock underlying the aquifer. The research shows that the NDA is still not in a state of dynamic equilibrium. The modeling instrument can be used for simulating future scenarios of SWI to provide a sustainable adaptation plan for groundwater resource.

3.2 Introduction

The Nile Delta ND is the most fertile land in Egypt, inducing extensive agricultural activities. Since the 1960s, these activities were primarily sustained by surface water from the Nile. In recent decades however, increasing water demand has brought the situation that the river can no longer provide sufficient amount of water. Consequently, the NDA is increasingly being exploited to obtain fresh groundwater. At the same time, these groundwater resources are being jeopardized by salt water intrusion (SWI), a common problem for coastal aquifers around the world, even more when groundwater extraction is substantial [1, 2]. In Egypt, SWI has already led to the salinization of groundwater extraction wells in coastal cities such as Alexandria [3]. This situation is expected to become worse; due to combined effects of climate change-induced sea level rise (SLR) [4], and excessive groundwater extraction for e.g., reclamation development projects [5], especially in El-Buhaira governorate (Figure 3.1). Overexploitation of fresh groundwater is also likely to occur in the ND as the rapidly growing population depends increasingly upon groundwater extraction for domestic water needs [6]. To counteract/manage/anticipate to these threats, a better understanding of the present hydrogeological conditions of the NDA is required: its geological condition, its groundwater dynamics, its relation with the surface water system and its current salinity concentration distribution.

In recent decades, different aspects of the ND have been investigated through research on geology [7, 8, 9], land subsidence and SLR [10, 11], geochemistry [12, 13], and SWI processes in the NDA and its hydrogeology [14]. Since the 1980's, researchers have used different numerical approaches to delineate fresh and saline groundwater interfaces. 2D models have been developed to simulate 2D cross-sections over the ND by [15, 16], while [17] used 2D models in a horizontal plane for eastern parts of the NDA.

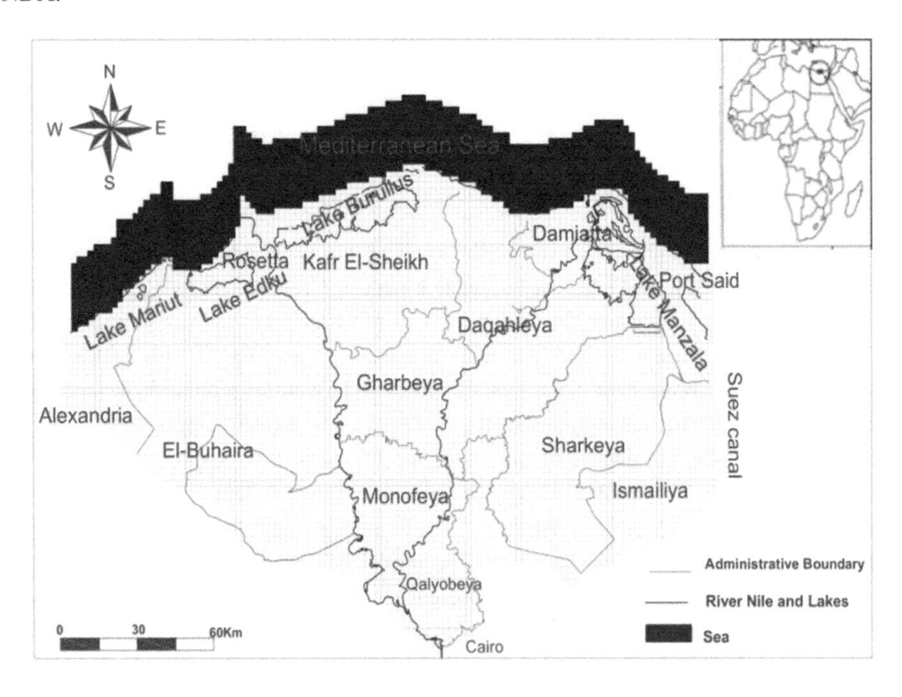

Figure 3.1. Location map of different governorates in the NDA

Modeling approaches for simulating 3D variable-density groundwater flow have recently shown rapid progress. A comprehensive overview of groundwater SWI situations, including relevant modeling approaches, is given in [2]. For the NDA, most research on quantifying variable-density groundwater flow processes has been carried out using 2D models, which cannot capture the full dynamics of the fresh groundwater-seawater interactions [17]. FEFLOW has been used by [18] to present the concept of equivalent fresh groundwater head in successive horizontal SWI simulations. Simulations in 2D were performed in four horizontal model layers representing different depths of the aquifer. Due to lack of sufficient data, several simplifying assumptions had to be made in that study regarding hydrological stresses, such as an average recharge rate over the entire ND. Later, [19] developed a 3D model in SEAWAT, using 28 vertical model layers, but most of the assumptions about hydrogeological stresses remained the same as in their previous work [18]. Followed by [20], who used SEAWAT to delineate the fresh and saline groundwater in the ND. They found that groundwater fluxes in the deep layers of the NDA are approaching the sea, which helps to retain the old brine in deeper zones. They used measured salinity concentrations as initial conditions and high longitudinal dispersivities (more than 1km, whereas normal values on delta scale are in order of tens of meters or lower). The acceptable bandwidth of dispersivitiy is comprehensively discussed in [21] as well as previously reported modeling cases of regional coastal aquifer systems [22, 23, 24]. Later, [25] investigated and proposed different strategies to protect the eastern NDA from SWI, but did not study the entire NDA. He found that the extraction of brackish groundwater provides very high reduction of SWI. However, the combination of decreasing extraction, increasing recharge, extracting brackish groundwater scenarios provide highest reduction of SWI.

Although previous studies have shown the applicability of variable-density groundwater flow modeling as a useful instrument, development of a reliable regional 3-D model for variable-density groundwater flow coupled with salt transport that can serve as tool for analyzing future scenarios and potential adaptation measures is still lacking. One obstacle in this development is the lack of sufficient amount of reliable hydrogeological data to be used for model setup, as well as for validating model outputs with monitored conditions of the groundwater resources in the NDA. In this chapter, we present the development of a 3D variable-density groundwater flow model and coupled with salt transport for the NDA using existing as well as new, more reliable data that have not been used before for similar purposes. The model has been developed to capture the situation in the year 2010 (representing present conditions), since for this year most data is available. Various types of reliable hydrogeological data have been collected from different departments and research institutions belonging to MWRI, as well as data from private sector organizations.

In addition to the contribution regarding the use of a rich dataset for the model development, the thesis also presents straightforward but innovative modeling procedure for determining the present 3D salinity concentration distribution (in the year 2010). First, the model is set up with structure and parametrization based on available data and expert judgment. Assuming that these are representative enough, the procedure tested nine models with different simulation periods, always starting with a completely fresh groundwater distribution in the entire NDA. The model that provides the best match between modeled and observed salinity concentration data in the year 2010 is then chosen as most reliable. Unlike some previous research aimed at developing full-fledged transient models e.g. [26, 27, 28], our procedure solely uses transient simulation models with the objective to capture best the salinity concentration distribution in the year 2010, without classical calibration approach of testing and adjusting different parameter values. The developed model can continue to be improved, but the obtained results indicate that it can already be used to assess the impacts of future SLR and groundwater extraction. From there, the effectiveness of adaptation and mitigation measures can be tested in future.

After this introduction, section two presents an overview of the physical settings of the NDA. It is followed by section three with a brief description of SEAWAT [29], the 3D code used in this study. This section includes the model setup, boundary conditions, hydrogeological parameters, hydrological stresses as well as the innovative procedure to determine the salinity concentration distribution. In section four, the salinity concentration distribution and the groundwater characteristics within the NDA are discussed, followed by a section conclusions and recommendations.

3.3 PHYSICAL SETTINGS

3.3.1 Study area

The modeled area shown in Figure 3.1 covers the ND, which has a triangular shape with an apex southwards near Cairo and with the Mediterranean Sea at its base. At the apex of the Delta, the Nile divides into two main branches: Damietta and Rosetta. The Suez Canal runs to the east of the ND, entering Lake Manzala in the northeast of the ND. Figure 3.2a shows the different elevations of the study area. The ND is the most populated region in Egypt, with an average population density of 1724 persons/km^2 [30]. It comprises a number of governorates with high economic and agricultural value.

3.3.2 Geology and aquifer characterization

The NDA is a semi-confined groundwater system, containing a huge groundwater reservoir [9]. It comprises Quaternary deposits that are classified in two main layers: the Holocene and the Pleistocene [31]. The Holocene deposit is composed of medium to fine-grained silt, with clay and peat in some regions [32]. It has a thickness of 50 m close to the sea [10] and vanishes towards the ND fringes in the south (Figure 3.2b). Groundwater exists in this geological layer at shallow depths ranging from 1.0 to 1.5 m below ground surface. The specific yield of the Holocene is very low, with low permeability of the clay and silt formations [31]. It is directly recharged by surface water infiltration from the River Nile, irrigation canals and drains, and also by excess of irrigation water. The lithology and the thickness variations of the Holocene affect the degree of the hydraulic connection between the surface water and groundwater systems.

The Pleistocene is the main aquifer of the entire NDA. It is a highly productive aquifer covering the entire ND. Its thickness varies from 200 m in the south up to 1000 m in the north (Figure 3.2c). The Pleistocene is composed of sand and gravel with occasional clay lenses. The sand and gravel are more common in the southern and middle regions of the ND. The clay lenses are more present in the north. Its underlying Pliocene formations are composed of shallow marine limestone and shale, characterized by low permeability [31].

3.4 METHODOLOGY AND MODEL SETUP

3.4.1 Code description

We based the regional numerical model of the ND on the conceptual groundwater aquifer system, geology and existing hydrological studies and data (Table 3.1). The computer code SEAWAT [29] is used to simulate the current groundwater and salinity concentration conditions. It simulates 3D variable-density groundwater flow, coupled with multispecies solute and heat transport; we used the conservative salt transport version. SWI is calculated by a variable-density groundwater flow model, [29]. We used PMWIN (Processing Modflow for Windows, Simcore Software 2010) for pre- and post-processing. For the governing equations, please consult the official SEAWAT manual for all relevant information (www.usgs.gov), as well as [29]. We selected SEAWAT as it had successfully been applied before in modeling numerous previous regional SWI studies.

SEAWAT, being based on the widely used MODFLOW and MT3DMS families, is available as free/open source, has a clear structure, and is supported by pre- and post-processors and additional tools that facilitate the development of models. SEAWAT supports different numerical methods for solving the solute transport equations such as TVD, FD, MOC, HMOC and MMOC; for more information see [33]. In this work, the advection term of the advection-dispersion equation was solved using the third-order Total-Variation Diminishing (TVD) method, based on the ULTIMATE algorithm (universal limiter for transient interpolation modeling of the advective transport equations) [33], while the generalized conjugate gradient (GCG) solver is was used for the non-advective terms.

Table 3.1. An overview of main data sets and its sources

Category	Variable	Source
Hydrogeology	Hydraulic conductivity Effective porosity Salinity concentration Groundwater extraction wells	MWRI [1] (RIGW [2]) Literature review CCC [3] GAEB [4] EGSM [5]
Climate	Meteorological data (rain)	Literature review
Land use	Agricultural zones Model boundaries	FAO [6] Literature review
Irrigation	Main canals and drains Discharge and water balance component	MWRI (CDS [7]) Literature review

[1] **MWRI**: Ministry of Water Resources &Irrigation.
[2] **RIGW**: Research Institute for Groundwater.
[3] **CCC:** Coca Cola Company in Egypt.
[4] **GAEB:** Geotechnical Authority for Educational Buildings
[5] **EGSM:** Egyptian Geological Survey and Mining
[6] **FAO:** Food &Agriculture organization
[7] **CDS:** Canal Distribution Sector

3.4.2 Model setup

Spatial discretization

The NDA model contains two main geological layers, the Holocene and the Pleistocene. The thickness of both aquifers is determined based on data collected from 2687 bore logs located in different governorates [34], combined with existing hydrogeological data [35]. The topography of the ND is quite flat, varying from 0 m mean sea level (MSL) at the sea coast to about 3 m MSL in the middle of the ND, and rising to elevations of about 40-50 m MSL at the southern fringes (Figure 3.2a). In the 3D model, the topography of the ground surface is specified using a topographic map of the ND on a scale of 1:50000 [36].

On the horizontal plane, the model is discretized using 100 rows and 150 columns, and using model cell sizes of about 2x2 km^2. The total area of the active model cells covered by the model is about 35.000 km^2. Vertically, the model is discretized with 21 model layers to simulate the fresh-brackish-saline groundwater dynamics on a regional scale. The top model layer represents the Holocene, followed by 19 model layers with equal thickness representing the Pleistocene. The last model layer represents the thin layer of the Pliocene formation that underlies the Pleistocene. This last model layer is introduced to make it possible to represent the dissolution of some salts and minerals from the underlying Pliocene formation (as will be further elaborated in the next section), see Figure 3.2d.

3.4.3 Boundary conditions

In the northern part of the model area, the Mediterranean Sea is represented as a constant head and constant salinity concentration boundary. The Mediterranean Sea is represented by 10 constant head and salinity concentration model cells (over a distance of 20 km) north from the present coastline (Figure 3.2a). We intentionally insert a significant portion offshore of the Mediterranean Sea into the model to have better representation of the boundary condition, as we assume the boundary condition affects the onshore groundwater system, especially when SLR is considered [37]. The decision to shift the constant head boundary condition 20 km offshore was determined by a rule of thumb: for a semi-confined aquifer the influence of a fixed boundary could be considered negligible at a distance of three times the leakage factor λ, being:

$$\lambda = \sqrt{\frac{D_1 K_h D_2}{K_v}} \tag{1}$$

Where D1 is the thickness of the Pleistocene [L], K_h is the horizontal hydraulic conductivity of the same coastal aquifer [LT^{-1}], D_2 is the thickness of the Holocene [L] and K_v is the vertical hydraulic conductivity of Holocene [LT^{-1}] [38]. Assuming average values of D_1=1000 m, D_2=25 m and Kv=0.025 m/day, K_h=0.25 m/day, λ = 7071 m. Proper boundary conditions are set when influences from that boundary do not propagate more than 5 percent into the semi-confined aquifer that is modeled. From the groundwater flow equation for semi-confined aquifer, which has an exponential form and includes the leakage factor λ, this condition is met when the distance between the boundary condition and the area of interest is set to 3 lambda (3λ) which is about 21 km (around 10 model cells) [37, 38].

A constant salinity concentration boundary is also assigned at all model cells representing the Mediterranean Sea (35 kg/m^3). The eastern boundary of the model corresponds to the Suez Canal and is considered a no-flow boundary. A no-flow boundary is also specified at the bottom of the Pliocene, which is the last model layer in vertical direction from ground surface. This model layer is introduced to represent the salinity concentration arising from the dissolution of marine deposits that underlie the Pleistocene aquifer (Figure 3.2d). The locations and actual concentration values representing these processes are determined based upon a number of isotope studies. Those studies have consistently agreed that the water type in the assigned locations is paleo-groundwater from ancient times [39, 40, 41, 42, 43, 44, 45]. Using radioactive 14C isotope, [46] proved the existence of paleo-groundwater in the southeastern region around the Ismailia canal, and attributed the salinity concentration of the groundwater to the dissolution of Pliocene minerals. Furthermore, [39, 40] proved, using tritium (3H), that the southwest region salinity concentration can be attributed to the terrestrial salts up-coning from the Pliocene formation. To consider this phenomenon in the model, we assigned to the bottom model layer in the southwest and southeast (locations detected in the isotopes studies) a constant concentration ranging between 1-4 kg/m^3 (Figure 3.2d).

3.4.4 Hydro geological parameters

Reports from a number of hydrogeological studies carried out in the ND were used to determine the hydrogeological parameters [8, 9]. The information required to identify and characterize the groundwater system was collected from the Research Institute of Groundwater (RIGW), the General Authority for Educational Buildings [34] and the

Coca Cola Company (Table 3.1). The primary source for the geological parameters and the lithological units of the NDA is a geotechnical database containing the results of a number of pumping tests carried out at different wells allocated throughout the ND governorates [34]. GAEB used the GWW (Ground Water for Windows) modeling tool to determine the parameters: hydraulic conductivity, safe yield and efficiency. All available data is collected and digitized, interpolated, and converted into formats required by the model. The pumping test data is used as primary source for specifying the values of horizontal hydraulic conductivity.

The first model layer (representing the Holocene) is specified to have a constant horizontal hydraulic conductivity of 0.25 m/day [5]. This layer consists predominantly of silt, mixed with some clay and sand lenses, but due to lack of data about spatially varying hydraulic conductivity, we have used spatially uniform value. For all the following nineteen model layers, representing the Pleistocene, horizontal hydraulic conductivity values vary from a minimum of about 15 m/day in the southwest, through a gradual increase from 50 m/day near the Mediterranean Sea to 130 m/day [34] in the middle of the ND (Figure 3.3). At some locations in the middle of the ND and in the southeast, horizontal hydraulic conductivity even reaches 150 m/day [8]. In the last model layer, representing the Pliocene, a horizontal hydraulic conductivity of 0.03 m/day [47] and 12.7 m/day [48] were assigned at their pre-determined locations in the southeast and southwest (shale and fractured marine limestone), respectively. The vertical hydraulic conductivity is assigned as 10 % of the horizontal hydraulic conductivity [49]. Spatially varying effective porosity for the Pleistocene is specified in the range of 12% near the Mediterranean Sea up to 28% in the south [34], with an average porosity value of 24% [8]. In the Holocene, the porosity is specified with a constant value of 40%. The porosity values are measured in the Faculty of Engineering labs, Cairo University, using bulk volume measurements, under the GAEB Project.

3.4.5 Hydrological stresses

Data on hydrological stresses on the NDA are collected from the MWRI and the national plans for groundwater of each governorate. In this subsection, a summary is given.

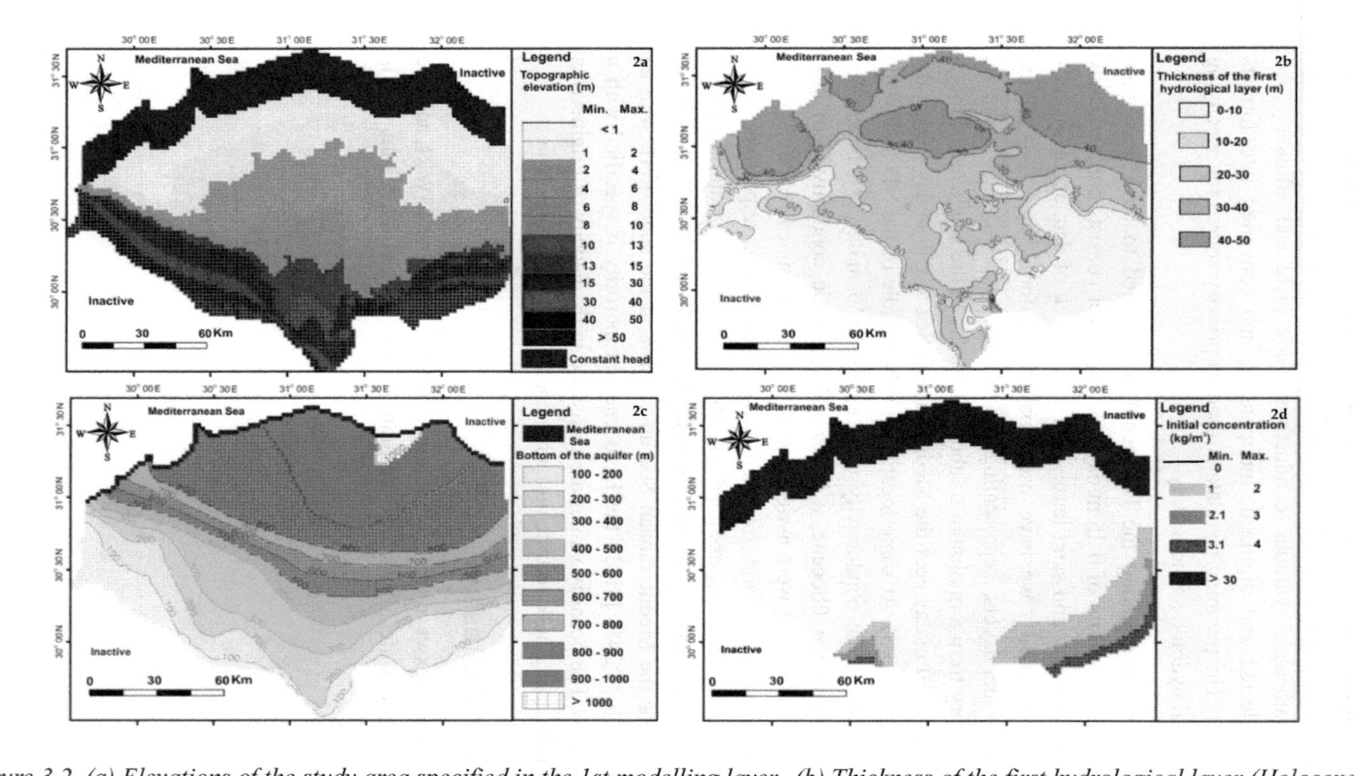

Figure 3.2. (a) Elevations of the study area specified in the 1st modelling layer. (b) Thickness of the first hydrological layer (Holocene). (c) The bottom of the second hydrological layer (Pleistocene). (d) Salinity concentration in kg/m^3 of the boundary condition in the deepest modelling layer, Modified from data used in [5]

Irrigation canals and drains

The irrigation and drainage networks in Egypt are extensive. We included the main canals and the second degree branch canals such as El-Nubariya, Bustan, Ismaillia, El-Baguriya and El-Mahmudiya, using the river package of MODFLOW/SEAWAT (Figure 3.3b). The water levels and different cross sections of the canals for the year 2010 are taken from the Canal Distribution Sector, MWRI [50]. All main and branch canals are divided into four or five segments, according to the measurement points available along the canal, for more accurate simulation (Figure 3.3b). Moreover, different water levels are also specified before and after weirs location along the canals. The average water level, bottom level and conductance are assigned for each segment [50]. To represent the small branch drains, a mesh representation of drains is specified using the drainage package, and is incorporated into the model (except for desert areas) in the first model layer, using a drain conductance of 2000 m^2/day.

Recharge

The main source of recharge in the NDA is infiltration from irrigation canals and excess agricultural irrigation water. Recharge from actual rainfall is very small, as precipitation varies from 200 mm/year along the Mediterranean Sea to 25 mm/year near Cairo [51]. In addition, with the high evaporation rate, the precipitation is negligible.

A land use map shows the ND to be covered principally by old traditionally cultivated land that is very fertile. Rice is the dominant crop in the coastal area. At the eastern and western fringes of the ND, new land with active reclamation projects is planned. The recharge values used in the model are based on data of [52], updated by [5]. They were obtained using hydrological budget calculations of the groundwater system and in situ measurements using infiltrometer rings in various ND governorates. The ND recharge rate is classified according to the irrigation type (basin, sprinkler, central pivot and drip irrigation), crop type (vegetables, barley, alfalfa, maize and fruits), soil type and structure, availability of artificial drainage and source of irrigation. Later, [5] used multi temporal Landsat imagery from 1992 TM (Thematic Mapper), 2005 TM and Google Earth Landsat 2008 of the Nile Region. She linked the image data with field studies, reports and topographic maps and determined the recharge rate for each governorate. Figure 3.3c shows the average recharge rates that are used in the model in different sectors of the ND. It shows that the value of the recharge rate varies from 0.01 mm/day in the desert to 1.1 mm/day in the south and even 1.9 mm/day in some western areas. Relatively high recharge rates are measured when irrigation is used in sandy soils: between 1.0 and 2.5 mm/day. In the northern part of the ND, the recharge rate is about 0.25 mm/day, increasing in the middle of the ND to 1 mm/day. Where drip irrigation is used, the recharge rates are much lower, ranging from 0.1-0.5 mm/day [5].

Salinity concentrations canals and recharge

Canal water in the ND is not completely fresh. Dissolved salts are introduced because of soil salinity and inefficient irrigation methods. The salinity concentration is incorporated into the model through the sink/source package. The salinity concentration data for the canals are obtained from a number of studies dealing with surface water quality [53, 54, 55]. Spatially varying salinity of rivers/canals and recharge is specified to the model to ensure that the groundwater entering the domain of the model has a proper salinity concentration. This approach has not been used in models developed in previous studies of the NDA.

Most canal water shows salinity concentrations of up to 0.3 kg/m^3. The salinity concentration of the main river branches, Rosetta and Damietta, start near Cairo with a value of 0.04 kg/m^3. Towards the north, canal water becomes more saline, especially near the Mediterranean, reaching up to 0.65 kg/m^3 where the soil salinity is very high (Figure 3.3d). Some regions show increased salinity due to irrigation methods e.g. flood irrigation for rice fields [5]. The salinity concentration in infiltration recharge ranges from 0.1 to 0.7 kg/m^3 [54].

Extraction wells

Groundwater extraction from wells is a critical activity in the ND. However, estimating the extraction amount is difficult. For instance, severe unauthorized extraction activities took place in the period from 1992-2008 [5]. In fact, overall groundwater extraction from the entire NDA in the year 2010 is estimated to have been at about 4.9x10^9 m^3/year compared to 3x10^9 m^3/year in 1992 [5]. The extraction wells used in the model represent both drinking water supply and irrigation wells. Due to the relatively coarse grid of this model, each model cell may represent the sum of a number of clustered wells in that model cell. The wells are distributed in different model layers according to their depth, which varies from between 45 m and 180 m below ground surface, meaning that most wells are placed in model layers six to fourteen. We collected extraction wells data from RIGW, the national plan for the year 2010 for each governorate [34], and the Coca Cola Company (Technical sheets and field reports 2011).

El Buhaira and Monofeya governorates in the western part of the ND have the maximum extraction rates, due to the uncontrolled increase in reclamation projects (Figure 3.4). The western region of the ND is characterized by rapid development in agriculture using groundwater. These extensive extraction rates are causing severe lowering of the water table and increased salinity concentration of the extracted groundwater. In addition, most of the wells north of the ND in Kafr El-Sheikh (30 wells) and Alexandria governorates have already been closed due to high salinity concentrations [5].

Solute transport characteristics

Advection and hydrodynamic dispersion processes for conservative salt transport are considered in the modeling. The longitudinal dispersivity is 1.0 m; the transversal horizontal dispersivity is 0.1 m and the transversal vertical dispersivity is 0.01 m. These relatively low values are assigned as acceptable estimates for the longitudinal dispersivities in the study of [21], as well as previously modeling cases of regional coastal aquifer systems e.g. [23].

3.4.6 Determining salinity concentration distribution of the year 2010

The constructed 3D model is being used to get a salinity concentration distribution that best fits the observed salinity concentration data. We designated the year 2010 as the reference year, as the most data are available for this year. Normally, two approaches are customarily used to determine the salinity concentration distribution: 1) using all existing observation wells for salinity concentration, and inter- and extrapolating geo-statistically the salinity values into a 3D salinity concentration distribution or 2) starting with a completely saline or fresh distribution and then simulating salt transport for a very long time until the groundwater system reaches dynamic equilibrium. The first approach can only be considered when enough data is available, which is seldom the case. With the second approach, there is more evidence now that especially huge groundwater systems like the NDA are not in a state of dynamic equilibrium, considering Palaeo-reconstructions and offshore fresh to brackish groundwater [56, 57, 58, 59, 60]. In our research, we introduced an approach that combines both options. The initial salinity concentration distribution of the model of the NDA is completely fresh with only saline concentration (35 kg/m^3) boundaries at the Mediterranean seaside. These boundary conditions might not always be feasible in other deltaic areas around the world. Our target is solely to determine the appropriate simulation period which provides the best match between the modeled and observed salinity concentration data.

In this approach, all simulations initially start without groundwater extraction. However, as groundwater extractions during the last 50 years have increased spectacularly, we took them into account assuming linear increase over this period. Thus, each simulation starts without groundwater extraction, but the last 50 years (assuming to represent 2060-2010) include linearly increasing groundwater extraction, from 0 to the value of $4.9x10^9$ m^3/year. All other hydrogeological stresses are kept constant in the model simulations. We have carried out model simulations for nine simulation periods varying from 200 to 2600 years, in order to determine the best match of the salinity distribution compared to the year 2010. This approach also allows checking whether the NDA has reached the state of dynamic equilibrium.

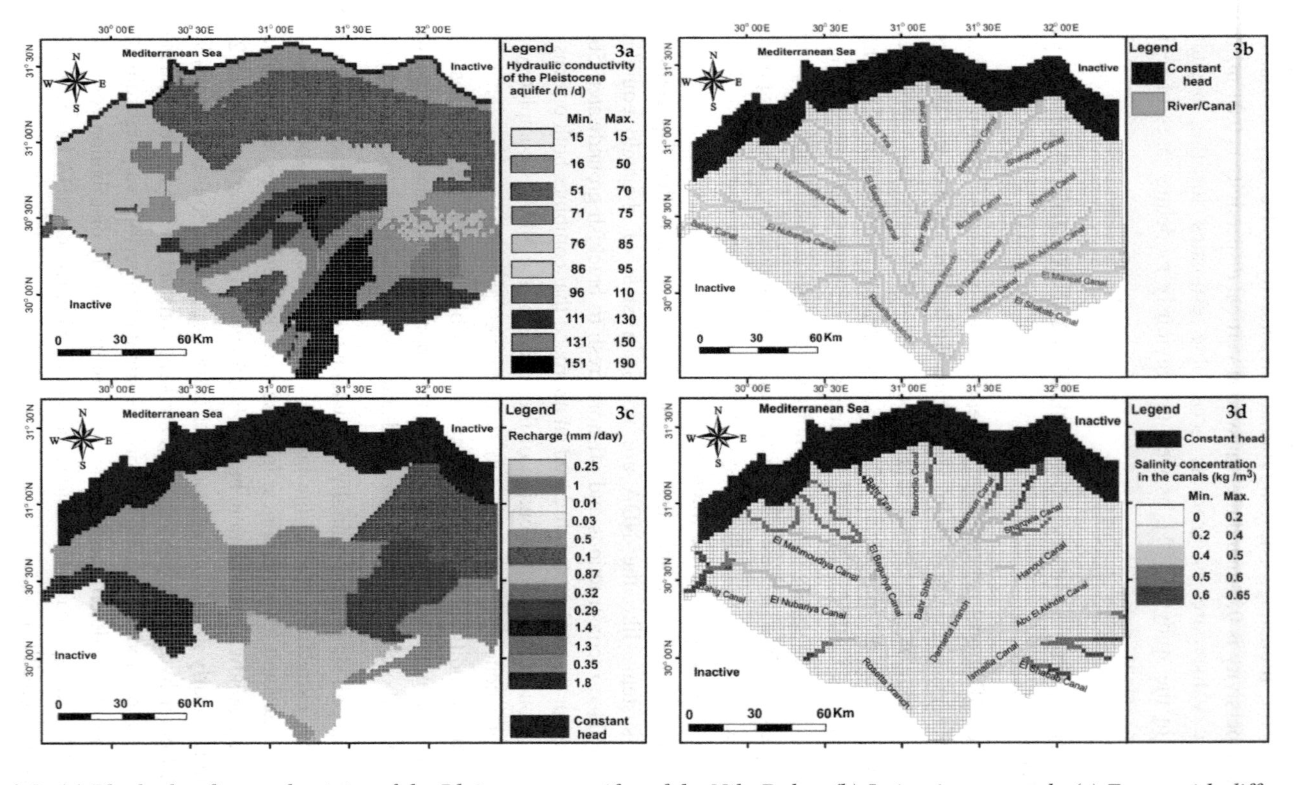

Figure 3.3. (a) The hydraulic conductivity of the Pleistocene aquifer of the Nile Delta. (b) Irrigation network. (c) Zones with different recharge rates in the Nile Delta. (d) The salinity concentration in kg/m³ the river package, Modified from data used in [5]

3.5 RESULTS AND DISCUSSION

All nine simulation periods showed as expected that the salinity front from the sea is rapidly progressing towards the southern NDA. Starting from an initial fresh groundwater distribution, the volume of saline groundwater increased with time in all simulated models. The larger the length of the simulation period, the longer the model is simulating SWI from the Mediterranean Sea, the more groundwater becomes salinized.

3.5.1 Comparing modeling results and observed salinity data for different simulation periods

To choose the best simulation period for the model, we compare the model results for salinity concentration with the observed salinity concentration data, and calculate the Root Mean Square Error (RMSE) for all nine models' simulations with different simulation periods. The comparison between modeled and observed salinity concentration is based upon observed concentration data from 2010 of 155 wells located throughout the NDA. We then determine as most reliable the model with the lowest RMSE value. Figure 3.5 shows the different RMSE values for all nine simulations' periods. The model with a simulation period of 800 years produces the lowest RMSE value. Hence, the salinity concentration distribution at the end of this simulation period with this model is selected.

Figure 3.6 shows some limited transient results (for the period that data is available) of modeled salinity concentrations results of six observations wells that are selected per regions as follows: two in the north, two in the middle and two in the south of the NDA. For each region, one point with best results is selected and another point with worst results. These results are presented for the period 1980-2010. A comparison between the values of the observed salinity concentration data with different simulation periods also indicates that the 800 years simulation period provides the best results. This is further confirmed with the results presented in Table 3.2 containing modeled and observed salinity concentrations for these six points in the year 2010.

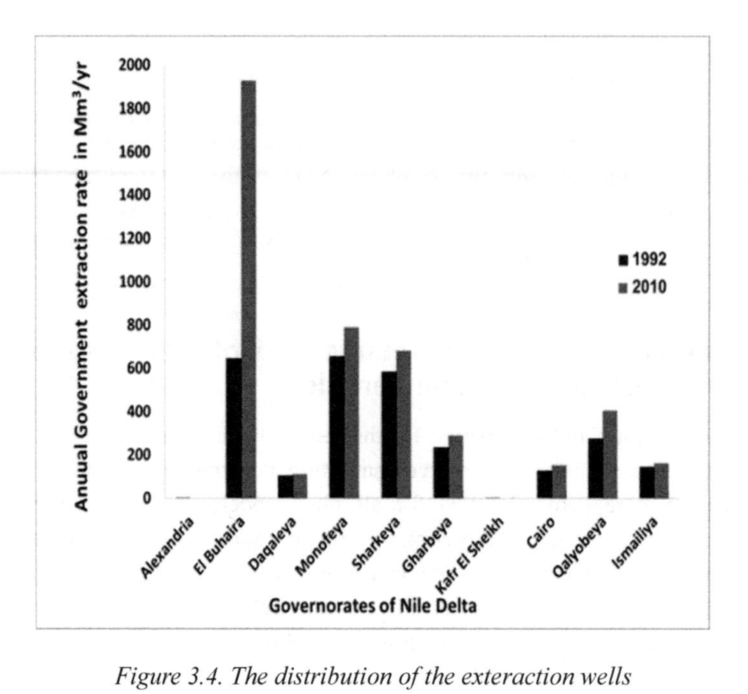

Figure 3.4. The distribution of the exteraction wells

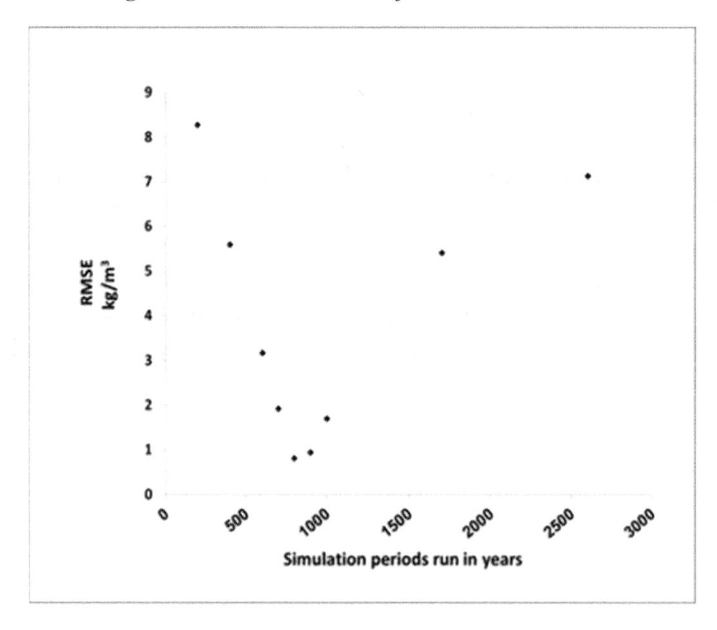

Figure 3.5. RMSE for different simulation periods w.r.t. salinity values

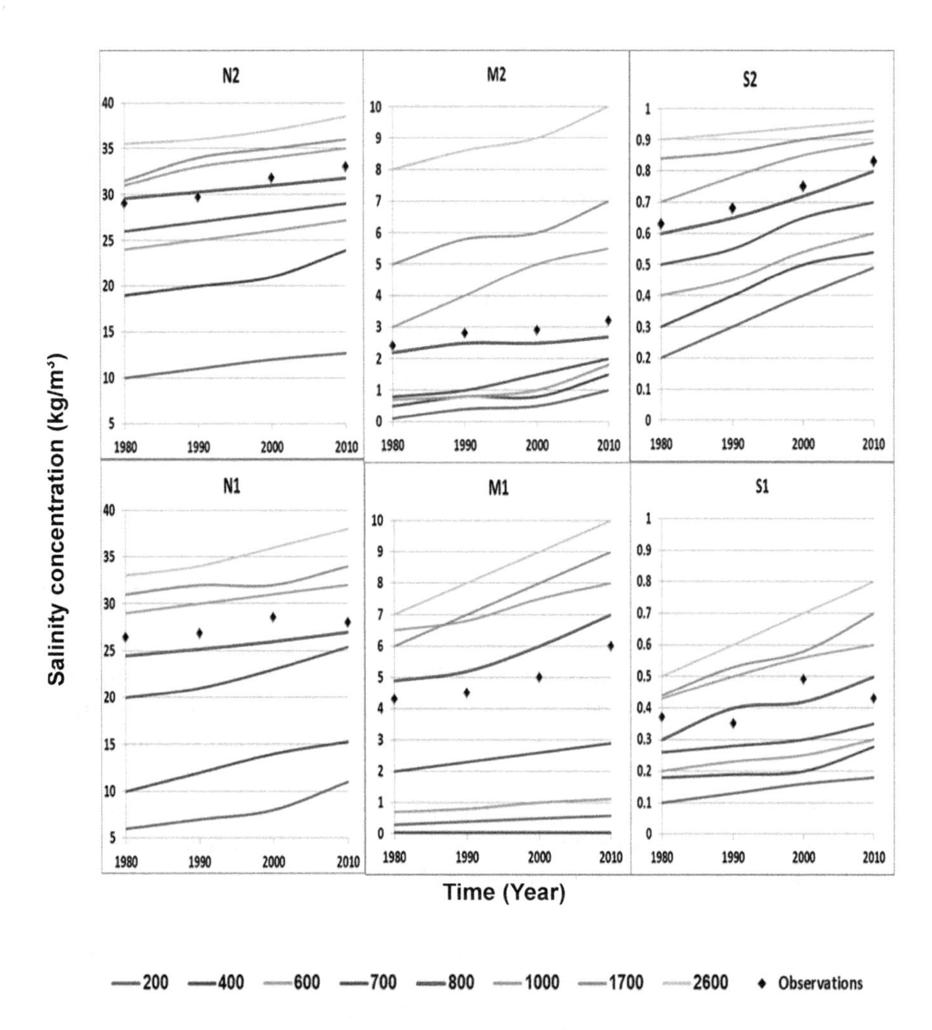

Figure 3.6. The salinity concentration values in kg/m³ for six points (S1, S2, M1, M2, N1 and N2) in the period (1980-2010) compared to their observed salinity concentration values for different simulation periods

Table 3.2. The salinity modeled concentrations values in kg/m^3 for six points (S1, S2, M1, M2, N1 and N2) in 2010 compared to their observed salinity concentration values for different simulation periods

Time interval of model simulation periods (years)	**Salinity modeled concentration in 2010**					
	S1	S2	M1	M2	N1	N2
200	0.18	0.49	0.59	1	11	12.68
400	0.28	0.54	0.05	1.5	15.3	23.9
600	0.3	0.6	1.116	1.8	22.9	27.15
700	0.35	0.7	2.9	2	25.4	29
800	0.5	0.8	7	2.7	27	31.8
1000	0.6	0.89	8	5.5	32	35
1700	0.7	0.93	9	7	34	36
2600	0.8	0.96	10	10	38	38.5
Observed salinity concentration	0.43	0.83	6	3.2	28	33

The simulations show that at present the NDA has not yet reached a state of dynamic equilibrium, as indicated by the gradient of the salinity concentration lines in Figure 3.6. This might be attributable to the substantial size of the groundwater system of the ND, which needs a very long time to achieve steady state conditions. On top, stresses such as the increasing rate of groundwater pumping cause additional non steady-state circumstances. The large size of the NDA in combination with Holocene characterized with lower permeability retards the interaction between the groundwater and the surface water system. In addition, the fact that salt transport is a slow process causes that the present state of the NDA is in non-equilibrium from a salinity concentration point of view. This condition has also been postulated by [19].

Figures 3.7a and b show the modeled versus observed salinity concentrations for the model with a simulation period of 800 years. The southern/middle part (a) and the northern part (b) are separated into two figures for clearer presentation. The absolute salinity concentration values are very small in the southern and middle parts of the NDA and large in the northern part. Figure 3.8 shows the frequency of the differences between modeled and observed concentration values. It shows that most of the frequency differences are between -0.75 and -0.5 kg/m^3. This implies that the differences are overall quite small, which increases the reliability of the results of the simulation.

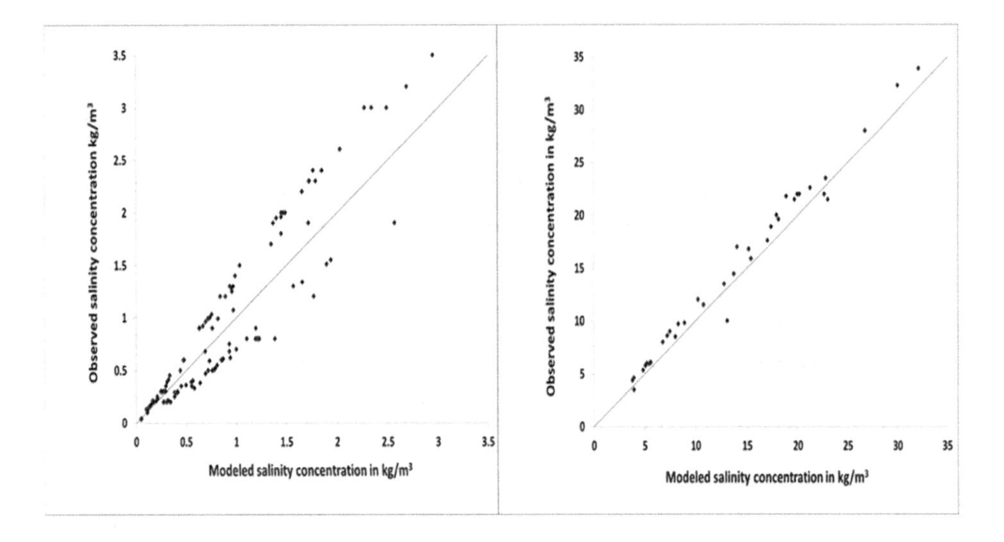

Figure 3.7. Modeled versus observed salinity concentrations in (a) The southern and the middle parts of the NDA (RMSE=0.8, Abs. Diff. =0.52, Stan Dev.=0.61kg/m³). (b) The northern parts of the NDA year 2010 (RMSE=0.2, Abs. Diff. =0.14, Stan Dev.=0.11kg/m³).

The RMSE values in the south are smaller than the RMSE values in the north, because the RMSE depends on the absolute values of differences between observed and modeled values. Modeled for fairer comparison of these values, Figure 3.9 shows the frequency of relative error (in percentage) of modeled salinity concentration. Although for most observations the relative error is relatively low (up to 30 %), it is clear that for about one third of observations the relative error is quite high (30%-45%). It should be noted, however, that when we further analyse these results together with results presented in Figures 3.7, 3.8 and 3.10 (see below), most of these high relative errors are for observations in the south and middle of NDA, where the absolute values of saline concentrations are very low. Figure 3.10 shows the relative error for observations in modeled different model layers. Note that observations from different layers are brought together to give better visualization of the distribution of the relative error. The dark blue dots represent observations belonging to the class of highest relative error (36% - 45%). Most of the dark blue dots in Figure 3.10 are scattered in the south, where the concentration of the infiltrated water varies significantly, and where absolute salinity values are very small, not exceeding 1 kg/m³. In the north (which is very sensitive to SWI), the absolute values of salinity go up to 30 kg/m³. In that part the relative error never exceeds 10% as it is shown in Figure 3.10 (represented by the distribution of yellow dots). This indicates that the large volume of reliable hydrogeological data used for the model set up results in low RMSE in this part of the model and in more reliable salinity distribution. Considering the regional scale of the NDA model and the wide range of salinity concentration values (0.045-35) kg/m^3, it can be concluded that these obtained relative errors are within acceptable range, but the model simulates the impact of SWI in the northern part better compared to the middle and the southern part.

The water budget terms have also been also analysed for the model. The huge irrigation network in the ND has had a great influence on the groundwater system. The main components of inflow are the net recharge due to excess irrigation water (around 5.3x10^9 m^3/year). Downward infiltration of surface water from the Nile and the main canals are minor in comparison: about 0.2 x10^{9}m^3/year. There is additional recharge in some regions due to groundwater recharge in reclamation areas, such as in El-Buhaira governorate in the western ND. The main outflow component is the extensive groundwater extraction (around 4.9x10^9 m^3/year), especially in the southern part of the eastern ND.

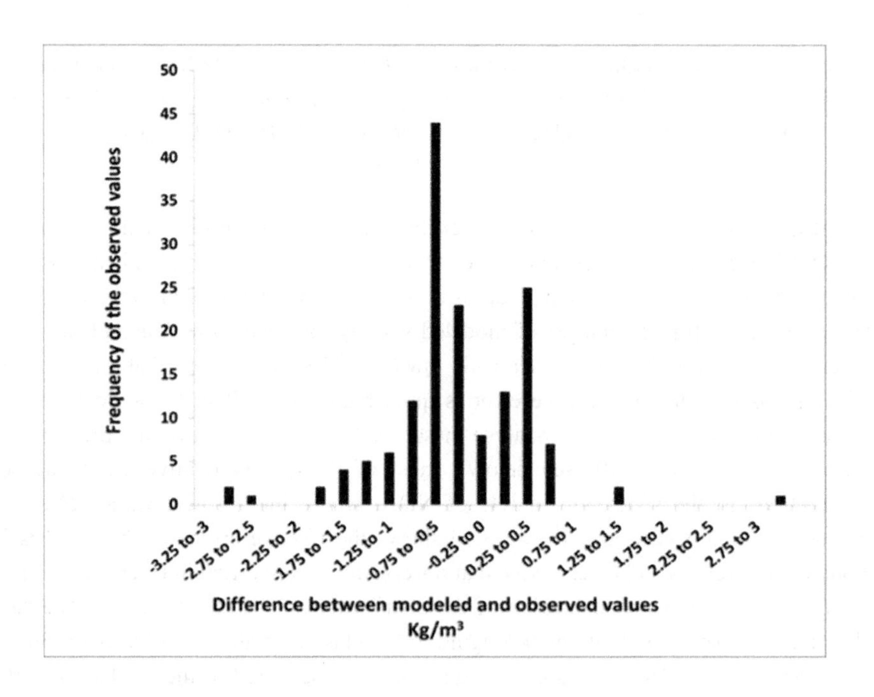

Figure 3.8. The histogram for the frequency of the modeled and observed concentrations

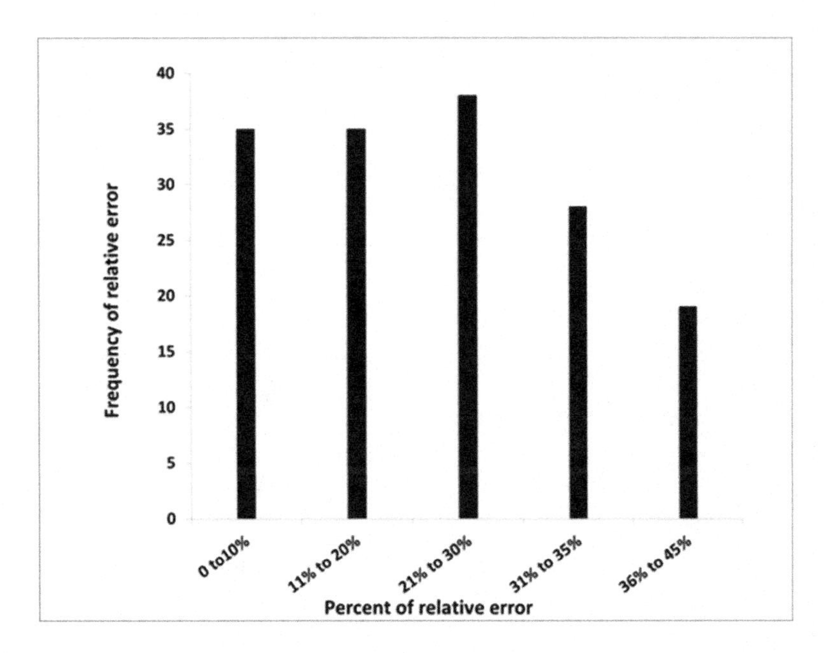

Figure 3.9. The frequency of relative error of modeled salinity concentrations with respect to observed concentrations. The total number of observations is 155

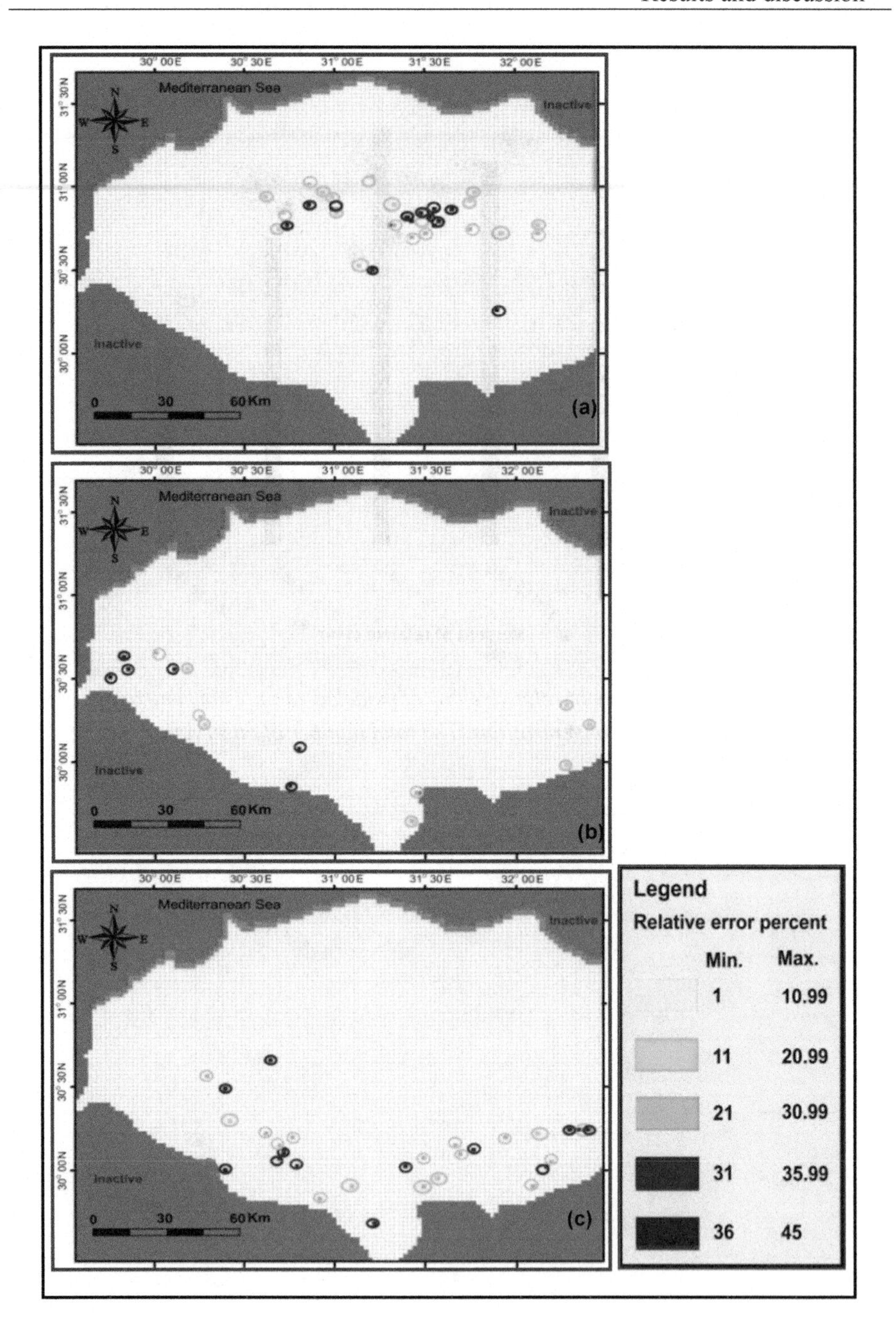

Figure 3.10. The relative error percent with respect to the modeled salinity concentration value. a. Layer 3, 4,5,6,7. b. Layer 8, 9, 10. c. Layer 11, 12, 13,14,15,16

3.5.2 Salinity concentration distribution

Figures 3.11a and b show the salinity distribution results for horizontal and vertical directions in the variable-density groundwater flow model with the simulation period of 800 years. Figure 3.11a presents the final salinity distribution in horizontal direction in model layers 1 and 21 at the end of the simulation period (representing the situation in the year 2010). We also presented how SWI increased with the progress of simulation time in a vertical cross-section across all model layers by using several selected time steps (Figure 3.11b). The new reliable data included in the model improved its performance and more reliable salinity distribution maps were obtained, when compared to previous studies. The south eastern region of the ND has been adequately represented and this is reflected clearly in the salinity distribution results and the RMSE values, unlike former studies that always showed this region as completely fresh [18].

These results initially indicate that the salinization process takes a long time. However, fingers of saline groundwater infiltrate rapidly into the fresh NDA (although with the model cell sizes only lumped vertical saline plumes can be simulated). Regarding the results obtained at the end of the simulation period, which represent the situation in the year 2010, it is clear that groundwater quality has already deteriorated. The salinity concentration varies from higher than 30 kg/m^3 to 10 kg/m^3 in the north. It decreases to about 1 kg/m^3 in Gharbeya governorate in the middle of the NDA (see for its location Figure 3.1 again). The salinity concentration in the southern region is very low, with values of less than 0.05 kg/m^3. The saline groundwater has spread in the north and northeast more widely than in the northwest, possibly because of the presence of a geological formation with higher hydraulic conductivities. Groundwater in the southern and middle regions is virtually fresh, being far distant from SWI introduced at the Mediterranean Sea boundary. All of the above results have been obtained due to the use of the 3D modelling approach, which when compared to previous 2D models [17] of the same area gives a clear overall salinity distribution in different layers of the NDA and captures the full dynamics of the fresh groundwater-seawater interactions.

The region in the east around the Sharkeya governorate contains light brackish groundwater (see salinity concentration contour of 1 kg/m^3 in Figures 3.11 a, b). This also occurs in the southwest around El- Buhaira governorate. Although the model results at individual points in this region are with larger errors, the overall higher salinity is still captured. As the infiltration rate has a lower value (0.01 mm/year) than other governorates, the source of saline groundwater in this region is probably coming from dissolution of deeper marine deposits, i.e. fractured marine limestone and shale, which was included in the last modelling layer. Isotope analysis and age dating tests discussed before have confirmed that water samples taken from those two regions contain paleo-water, probably from the underlying Pliocene formation [44]. This is however, combined with severe abstraction due to extensive reclamation activities in El-Buhaira governorate increases the salinity concentration of water drawn up from the deeper parts of the NDA.

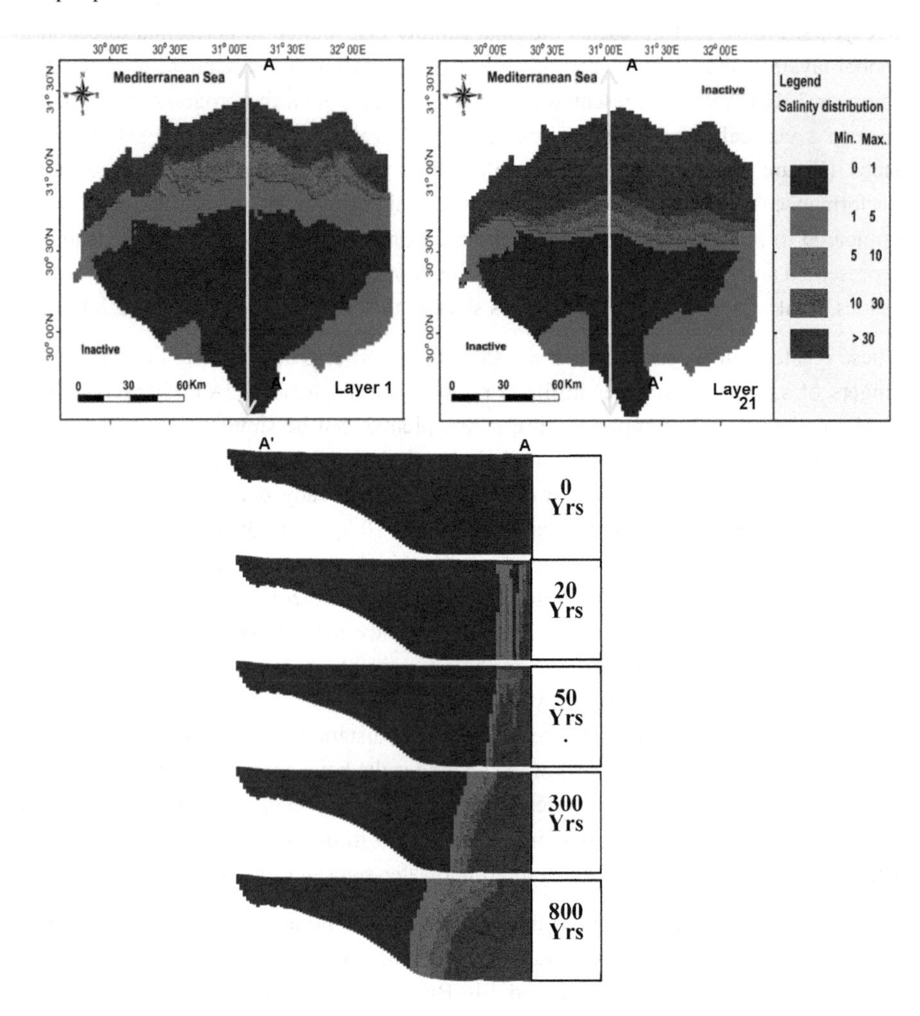

Figure 3.11. (a) The salinity concentration distribution in kg/m³ in first and last layers for 2010AD (last time step of the model with simulation period of 800 years) (b) Salinity concentration distribution along the cross section of the NDA versus time (A-A')

3.6 CONCLUSIONS AND RECOMMENDATIONS

Earlier research on the fresh and saline groundwater system in the NDA under the pressure of SWI using numerical models has mainly focused on 2D simulations, which does not fully imply the complex nature of the NDA in three dimensions. 3D numerical modelling studies have been executed before, but lacked sufficient hydrogeological data or with different parametrization. In this research, a fully 3D variable-density groundwater flow model is constructed to simulate the current situation of salinization in the NDA. This NDA model incorporates hydrogeological data from many sources. An innovative procedure is performed without classical calibration. It uses physically sound model parameters, starting from completely fresh ground water in the NDA and sets up several different simulations periods to obtain the best salinity concentration distribution in the year of 2010. After simulation, the 3D fresh-saline distribution that fits best the observed salinity data is chosen. Future studies need to address the influence of different parametrization on modeling results in combination with the proposed approach here. This is beyond the scope of the current paper but could be done within existing frameworks for sensitivity and uncertainty analyses of groundwater models. The developed 3D model represents well the salinity distribution in the NDA. The comparison between numerous (spatially varying) groundwater salinity concentration observations and modeled results shows that the model is capable of representing the current salinity distribution in the NDA, taking the year 2010 as the reference year. Therefore, we believe the model can be used with increasing confidence for future predictions. The model can be considered reliable but that there are areas in the south where the error is always quite high. When performing forecasting simulations, it may be useful to further improve the model by means of a validation process based on new data collected. The validation process with data collected after 2010 could also confirm the innovative procedure used.

The model results indicate that specific regions in the east (Sharkeya governorate) and southwest (El-Buhaira governorate) likely suffer from salinization due both to dissolution of marine deposits and excessive extraction. Both governorates have great economic agricultural value in Egypt, especially in commodities export. Further study on the fresh water needs and methods of extraction in these governorates might affect both conservation efforts and economic development planning with a view to conservation of the fresh groundwater resources in the NDA.

The model serves as a tool for assessing areas of the groundwater system vulnerable to salinization due to combined stresses, such as increased groundwater extraction, different irrigation practices, and SLR. It can be used to track the movement of fresh, brackish and saline groundwater in the NDA, and for testing future adaptation /mitigation measures for the NDA. Such effective use of the model, however, critically depends on data availability. More salinity concentration data are especially needed, especially from greater depths. Future monitoring campaigns must also better control the salinity sampling process due to its effect on observed sample biases. Multi-level monitoring samplers should be used, especially when lateral SWI and up-coning processes are taking place. These aspects have not been sufficiently controlled past sampling procedures, which may affect modelling results and consequently future predictions. Although collecting reliable monitoring data is costly, monitoring programs are strongly recommended for deep wells, as fresh groundwater resources are of eminent importance for Egypt. Fortunately, faster and more reliable salinity monitoring techniques have become more efficient in the last decade. The Airborne Electromagnetic (AEM) geophysical methods look especially promising [61, 62, 63, 64, 65]. The availability of more data will be a keystone for further development in the potential of this model, and will extend the horizon of research in the NDA for different water perspectives.

Finally, the experiences from the developed model, and from the procedure proposed for obtaining more reliable model setup, could also be used and tested in aquifer system in different deltas of the world where groundwater resources are deteriorating due to SWI. In spite of differences in geometry and their hydrological parameters, most deltaic areas face the similar development and climate stresses.

REFERENCES

1 Custodio, E. Aquifer overexploitation: what does it mean? Hydrogeology J. 10, 2, 254–277, doi 10.1007/s10040-002-0188-6, 2002.

2 Werner, A.D., Bakker, M., Post, V.E.A., Vandenbohede, A., Lu., C., Ataie-Ashtiani, B., Simmons, C.T., Barry, D.A. Seawater intrusion processes, investigation and management: Recent advances and future challenges. *Adv. Water Res. J.*, 51, 3-26, 2013.

3 El Raey, M., Nasr, S., Frihy, O., Desouki, S., and Dewidar, Kh. Potential impacts of accelerated sea level rise on Alexandria governorate, Egypt, *Coastal Research. J.* 14, 190–204, 1995.

4 Oppenheimer, M., B.C. Glavovic, J. Hinkel, R. van de Wal, A.K. Magnan, A. Abd-Elgawad, R. Cai, M. Cifuentes-Jara, R.M. DeConto, T. Ghosh, J. Hay, F. Isla, B. Marzeion, B. Meyssignac, and Z. Sebesvari,: Sea level rise and implications for low-lying islands, coasts and communities. In: IPCC special report on the ocean and cryosphere in a changing climate, 2019.

5 Morsy, W.S. Environmental management to groundwater resources for Nile Delta region. Ph.D. thesis, Fac. of Eng., Cairo Uni., Egypt, (www.cu.edu), 266, 2009.

6 Mabrouk, M.B., Jonoski, A., Solomatine, D., Uhlenbrook, S. A review of seawater intrusion in the Nile Delta groundwater system—The basis for assessing impacts due to climate changes, sea level rise and water resources development. *Nile Water Sci. Eng. J.* 2017, 10, 46–61, ISSN 2090-0953, 2017.

7 Pennington, B.T., Sturt, F., Wilson, P., Rowland, J., Brown, A.G. The fluvial evolution of the Holocene Nile Delta. *Quaternary Sci. Reviews J.* 170, 212–231, doi 10.1016/2017.06.017, 2017.

8 Dahab, K. Hydrogeological evaluation of the Nile Delta after High Dam construction. Ph.D. thesis, Fac. of Sci., Menoufia Uni., Egypt, 1993.

9 Farid, M.S. Management of groundwater system in the Nile Delta. Ph.D. thesis, Fac. of Eng., Cairo Uni, 1980.

10 Stanley, J., Clemente, P.L. Increased land subsidence and sea level rise are submerging Egypt's Nile Delta coastal margin. *Geological Society of America J.* 27, 5, doi10.1130/312A.1, 2017.

11 Stanley, D.J. Subsidence in the northeastern Nile Delta: rapid rates, possible causes, and consequences. *Sci. J.* 240, 4851, 497-500, 1988.

12 Geriesh, M.H., Balke, K.D., El-Rayes, A.E., Mansour, B.M. Implications of climate change on the groundwater flow regime and geochemistry of the Nile Delta, Egypt. *Coastal Conservation J.* 19, 4, 589–608, doi 10.1007/s11852-015-0409-5, 2015.

13 Abd-ElMoati, M.A.R. and El-Sammak, A.A. Man-made impact on the geochemistry of the Nile Delta Lakes. A study of metals concentrations in sediments. *Water, Air, and Soil Pollution J.* 97, 3-4, 413, 1997.

14 Abd-Elhamid, H.F., Javadi, A.A., Abd-Elaty, I., Sherif, M.M. Simulation of seawater intrusion in the Nile Delta aquifer under the conditions of climate change. *Hydrology Research J.* 47, 6, 1198-1210. doi 10.2166/nh.2016.157, 2016.

15 Darwish, M.M. Effect of probable hydrological changes on the Nile Delta aquifer system. Ph.D. thesis, Cairo Uni. Egypt, 1994.

16 Sherif, M.M., Singh, V.P. Effect of climate change on seawater intrusion in coastal aquifers. *Hydrological Processes J.* 13, 1277–1287, 1999.

17 Nosair, A. Climatic change and their impacts on groundwater occurrence in eastern part in Nile Delta. Msc. thesis, Fac. of Sci., Zagazig Uni., Egypt, 228, 2011.

18 Sherif, M.M., Sefelnasr, A., Javad, A. incorporating the concept of equivalent freshwater head in successive horizontal simulations of seawater intrusion in the Nile Delta aquifer, Egypt. *Hydrogeology J.* 464, 186-198, doi 10.1016/2012.07.007, 2012.

19 Sefelnasr, A., Sherif, M.M. Impacts of seawater rise on seawater intrusion in the Nile Delta aquifer, Egypt. *Groundwater J.* 52, 2, 264-276, doi 10.1111/gwat.12058, 2014.

20 Nofal, E.R., Amer, M.A., El-Didy, S.M., Fekry, A.M. Delineation and modeling of seawater intrusion into the Nile Delta Aquifer: a new perspective. *Water Sci. J.* 29, 6, 156–166, 2015.

21 Gelhar, L.W., Welty, C., Rehfeldt, K.R. A Critical Review of Data on Field-Scale Dispersion in Aquifers. *Water Resources Research J.* 28, 7, 1955-1974, doi 0043·1397/92/92wr-00607sos.00,1992.

22 Werner, A.D., Gallagher, M.R. Characterization of seawater intrusion in the Pioneer Valley, Australia using hydrochemistry and three-dimensional numerical modelling. *Hydrogeology J.* 14, 1452–1469, 2006.

23 Oude Essink, G.H.P., Van Baaren, E. S., De Louw, P.G.B. Effects of climate change on coastal groundwater systems: a modeling study in the Netherlands. *Water Resources Research J.* 46, 10, doi 10.1029/2009WR008719, 2010.

24 Zeghici, R.M., Oude Essink, G.H.P., Hartog, N., Sommer, W. Integrated assessment of variable density–viscosity groundwater flow for a high temperature mono-well aquifer thermal energy storage (HT-ATES) system in a geothermal reservoir. *Geothermic J.* 55, 58–68, doi10.1016/2014.12.006, 2015.

25 Abd-Elhamid, H.F. Investigation and control of seawater intrusion in the Eastern Nile Delta aquifer considering climate change. Water Sci. and Technology, *Water Supply J.* 17, 2, 311-323, doi 10.2166/ws.2016.129, 2017.

26 Yang, J., Graf, T., Ptak, T. Impact of climate change on freshwater resources in a heterogeneous coastal aquifer of Bremerhaven, Germany: A three-dimensional modeling study. *Contaminated Hydrology J.* 177,107–121. Doi 10.1016/2015.03.014, 2015.

27 Gingerich, S.B., Voss, C.I. Three-dimensional variable-density flow simulation of a coastal aquifer in southern Oahu, Hawaii, USA. *Hydrogeology J.* 13, 2, 436–450, doi 10.1007/s10040-004-0371-z, 2007.

28 Schulz, S., Walther, M., Michelsen, N., Rausch, R., Dirks, H., Al-Saud, M., Merz, R., Kolditz, O., Schüth, C. Improving large-scale groundwater models by considering fossil gradients. *Adv. Water Res.J.* 103, 32-43 doi10.1016/.2017.02.010, 2017.

29 Langevin, C.D., Thorne, D.T., Dausman, J.R., Sukop, A.M., Guo, W. SEAWAT Version 4: A Computer Program for Simulation of Multi-Species Solute and Heat Transport: U.S. Geological Survey Techniques and Methods Book, 6-A22, 39,2008.

30 CAPMAS, Central Agency for Public Mobilization and Statistics Egypt (www.capmas.gov.eg), 2017.

31 Sestini, G. Nile Delta: A review of depositional environments and geological history. *Geological Society J.* London, Special Publications, 41, 1, 99-127, doi10.1144/GSL.SP.1989.041.01.09, 1989.

32 Said, R. The geologic evolution of the River Nile. *Springer Sci. and Business Media J.* New York, 151, doi 10.1007/978-1-4612-5841-4,1981.

33 Zheng, C., Wang, P.P. MT3DMS: A Modular Three- Dimensional Multispecies Transport Model for Simulation of Advection, Dispersion and Chemical Reactions of Contaminants in Groundwater Systems; Documentation and User's Guide., Strategic Environmental Research and Development Program,1999.

34 GAEB. Geotechnical encyclopedia of Egypt. Book, 1288, I.S.B.N.977-6079-23-7, 2011.

35 RIGW. Research Inst. for Groundwater, Ministry of water resources and irrigation. Hydrogeology Map of the Nile Delta, Scale 1: 50000000, 1st Ed, 1992.

36 EGSA. Egyptian General Survey and Mining, Topographical Map cover Nile Delta, scale 1:2.000.000, 1997.

37 Walther, M., Graf, T., Kolditz, O., Liedl, R., Post, V. How significant is the slope of the sea-side boundary for modelling seawater intrusion in coastal aquifers? *Hydrology J.* 551, 648–659, doi10.1016/J.Hydrol.2017.02.031,2017.

38 Huisman, L. Groundwater Recovery. Besleys Books, United Kingdom, 336, ISBN 10: 0333098706, 1972.

39 Aly, A.I.M., Simpson, H.J., Hamza, M.S., White, J.W.C., Nada, A., Awad, M.A. Use of environmental isotopes in the quantification of the water budget of the Nile Delta, Egypt. *Isotope and Radiation Research J.* 21, 1, 31-40, ISSN 0021-1907, 1991.

40 Awad, M.A., Nada, A.A., Aly, A.I.M., Farid, M.S. Tritium content of groundwater aquifer in western Nile Delta. *Isotopes in Env.and Health Studies J.* 28, 2, 167-173, doi 10.1080/10256019308682866, 1993.

41 Awad, M.A., Farid, M.S., Hamza, M.S. Studies on the recharge of the aquifer systems in the southern portion of the Nile Delta. *Isotope and Radiation Research J.* 26, 1, 21-25, ISSN 0021-1907, 1994.

42 Awad, M.A., Sadek, M.A., Salem, W.M. Use of mass balance and statistical correlation for geochemical and isotopic investigation of the groundwater in the quaternary aquifer of the Nile Delta, Egypt. *Arab journal of Nuclear Sci. and App. J.*32, 1, 43-58, ISSN 1110-0451,1999.

43 Hamza, M.S., Aly, A.I.M., Swailem, F.M., Nada, A. Environmental stable isotopes indicate possible sources of groundwater recharge in eastern Nile Delta. *Isotopes and Radiation Research J.*1987, 19, 1, 7-14, ISSN 0021-1907, 1987.

44 Salem, W.M., Sadek, M.A. Chemical and isotopic signature for mapping zones of seawater intrusion and residual saline pockets in coastal aquifers of the Egyptian Nile Delta. *Arab Nuclear Sci. and Application J.* 39, 3, 111-127, ISSN 1110-0451, 2006.

45 Shohaib, R. A study of environmental isotopes and trace elements abundance in Egyptian water. Ph.D. thesis, Fac. of Sci., Cairo University, Egypt, 1980.

46 Salem, W. M. M. Application of environmental isotopes and hydrochemical modelling to study groundwater in Tenth of Ramadan city, Egypt. *Isotope and Radiation Research J.* 33, 2, 243-259, ISSN 0021-1907, 2001.

47 Webster, D.S., Procter J.R., Marine, J.W. Two-well tracer test in fractured crystalline rock. USGS, Water Supply J. 1544, 1970.

48 Segol, G., Pinder, G.F. Transient simulation of saltwater intrusion in southeastern Florida. *Water Resources Research J.* 12, 1, 65-70, doi 10.1029/WR012i001p00065, 1976.

49 Bear, J. Hydraulics of groundwater, Dover Publications, Courier corporation, 592, ISBN 0-486-45355-3, 2007.

50 CDS. Canal Distribution Sector, Ministry of Water Resources and Irrigation in Egypt, National yearly report, Vol. 10, 2010.

51 El-Sadek, A., M. El Kahloun, and P. Meire. Ecohydrology for integrated water resources management in the Nile Basin. *Ecohydrology and Hydrobiology J.* 2008, 8, 2–4, 77–84, 2008.

52 RIGW/IWACO, Environmental management of groundwater resources (EMGR), Final technical report TN/70.0067/WQM/97/20, Kanater El-Khairia, Egypt, 1999.

53 Azab, A. Surface water quality management in irrigated watersheds. Ph.D. thesis, Delft Uni. of Tech., UNESCO-IHE, 240, ISBN978-0-415-62115-1, 2012.

54 Elewa, H.H., El Nahry, H. Hydro-environmental status and soil management of the River Nile Delta, Egypt. Springer-Verlag, *Env. Geology J.* 2009, 57, 4, 759–774, 2009.

55 Taha, A.A., El-Mahmoudi, A.S., El-Haddad, I.M. Pollution sources and related environmental impacts in the new communities southeast Nile Delta, Egypt. *Emirates Journal for Eng. Research J.* 9, 1, 35-49, 2004.

56 Delsman, J.R., Hu-a-ng K.R.M., Vos. P.C.C., De Louw, P.G.B., Oude Essink, G.H.P, Stuyfzand. P.J., Bierkens. M.F.P. Paleo-modeling of coastal saltwater intrusion during the Holocene: An application to the Netherlands. *Hydrology and Earth System Sci. J.* 18, 10, 3891–3905 doi 10.5194/hess-18-3891, 2014.

57 Meisler, H., Leahy, P.P., Knobel, L.L. Effect of eustatic sea level changes on saltwater-freshwater in the northern Atlantic Coastal Plain. In USGS Water Supply Paper: 2255U.S. *Geological Survey J.* 28-34, 1984.

58 Misut, P.E., Voss, C.I. Freshwater–saltwater transition zone movement during aquifer storage and recovery cycles in Brooklyn and Queens, New York City, USA. *Hydrology J.* 337, 1-2, 87–103 doi 10.1016/2007.01.035, 2017.

59 Post, V.E.A., Groen, J., Kooi, H., Person, M., Ge.S., Edmunds, W.M. Offshore fresh groundwater reserves as a global phenomenon. *Nat. J.* 504, 7478, 71–8 doi 10.1038/nature12858, 2013.

60 Larsen, F., Tran, L.V., Van Hoang, H., Tran, L.T., Christiansen, A.V., Pham, N.Q. Groundwater salinity influenced by Holocene seawater trapped in incised valleys in the Red River delta plain. *Nat. Geosci. J.* 10, 376–381. doi10.1038/ngeo2938, 2017.

61 Delsman, J., Van Baaren, E.S., Siemon, B., Dabekaussen, W., Karaoulis, M.C., Pauw, P., Vermaas, T., Bootsma, H., De Louw, P.G.B., Gunnink, J.L., Dubelaar, W., Menkovic, A., Steuer, A., Meyer, U., Revil, A., Oude Essink, G.H.P. Large-scale, probabilistic salinity mapping using airborne electromagnetics for groundwater management in Zeeland, the Netherlands. *Env. Res. J.* doi10.1088/1748-9326/aad19e, 2018.

62 Faneca Sànchez, M., Gunnink, J.L., Van Baaren, E.S., Oude Essink, G.H.P., Auken, E., Elderhorst, W., De Louw, P.G.B. Modelling climate change effects on a Dutch coastal groundwater system using airborne electromagnetic measurements. *Hydrology and Earth System Sci. J.* 2012, 16, 12, 4499–4516, doi 10.5194/hess-16-4499, 2012.

63 Jorgensen, F., Scheer, W., Thomsen, S., Sonnenborg, T.O., Hinsby, K., Wiederhold, H., Schamper, C., Burschil, T., Roth, B., Kirsch, R. Transboundary geophysical mapping of geological elements and salinity distribution critical for the assessment of future sea water intrusion in response to sea level rise. *Hydrology and Earth System Sci. J.*16, 7, 1845–1862, doi 10.5194/hess-16-1845, 2012.

64 Rasmussen, P., Sonnenborg, T.O., Goncear, G., Hinsby, K. Assessing impacts of climate change, sea-level rise, and drainage canals on saltwater intrusion to coastal aquifer. *Hydrology and Earth System Sci. J.* 17, 1, 421–443 doi 10.5194/hess-17-421, 2013.

65 Siemon, B., Christiansen, A.V., Auken, E. A review of helicopter-borne electromagnetic methods for groundwater exploration. Near *Surface Geophysics J.* 2009, 7, 5, 629–646, 2009.

4 IMPACTS OF SLR AND GROUNDWATER EXTRACTION SCENARIOS ON FRESH GROUNDWATER RESOURCES IN THE NILE DELTA GOVERNORATES, EGYPT

The content of this chapter is a full reproduction of the published article:

Mabrouk, M.; Jonoski, A; Oude Essink, G.H.P.; Uhlenbrook, S. Impacts of Sea Level Rise and Groundwater Extraction Scenarios on Fresh Groundwater Resources in the Nile Delta Governorates, Egypt. *Water* J. 10, 1690, 2018.

4.1 ABSTRACT

As Egypt's population increases, the demand for fresh groundwater extraction will intensify. Consequently, the groundwater quality will deteriorate, including an increase in salinization. On the other hand, salinization caused by saltwater intrusion (SWI) in the coastal Nile Delta Aquifer (NDA) is also threatening the groundwater resources. The aim of this chapter is to assess the situation in 2010 (since this is when most data is sufficiently available) regarding the available fresh groundwater resources and to evaluate future salinization in the NDA using a 3D variable-density groundwater flow model coupled with salt transport that was developed with SEAWAT. This is achieved by examining six future scenarios that combine two driving forces: increased extraction and sea level rise (SLR). Given the prognosis of the intergovernmental panel (IPCC) on CC, the scenarios are used to assess the impact of groundwater extraction versus SLR on the saltwater intrusion in the Nile Delta (ND) and evaluate their contributions to increased groundwater salinization. The results show that groundwater extraction has a greater impact on salinization of the NDA than SLR, while the two factors combined cause the largest reduction of available fresh groundwater resources. The significant findings of this research are the determination of the groundwater volumes of fresh water, brackish, light brackish and saline water in the NDA as a whole and in each governorate and the identification of the governorates that are most vulnerable to salinization. It is highly recommended that the results of this analysis are considered in future mitigation and/or adaptation plans.

4.2 INTRODUCTION

The ND in Egypt is occupied by the most populated governorates in Egypt. About 60% of Egypt's population lives in the ND region [1]. Agriculture activities are dominant in the region due to the nature of the soil and the presence of an irrigation system [2]. The NDA is a vast aquifer located between Cairo and the Mediterranean Sea [3]. The NDA was formed by Quaternary deposits with a wide variety of hydrogeological characteristics and spatially varying salinity levels [4]. These deposits represent different aggradation and degradation phases that were usually accompanied by sea level changes [5]. Recent years have brought scientific evidence of increased groundwater salinization in NDA, predominantly driven by increased groundwater extraction [4].

Salinity in groundwater is a major quality hazard that limits its usage and affects the productivity of agricultural areas that depend on irrigation from groundwater wells [6]. The nature and properties of salinity were reviewed with a view to its management in [7]. In coastal aquifers, such as the NDA, the salinity in groundwater is influenced by human interventions through excessive groundwater extraction, while SWI induced by sea-level rise SLR is also anticipated [8].

In north Kuwait, a freshwater aquifer polluted by saline seawater was modelled using a 3D numerical model [9]. The researchers declared that solute transport modeling has become a significant tool for analysing the groundwater quality, as it can provide insight into past and present behaviour and predict water quality management scenarios. It was also highlighted that CC is likely to lead to multiple stresses in the groundwater sector [10]. This research emphasized the need for the development of management models that simulate SWI in aquifers and the assessment of future compound groundwater challenges. Open challenges and uncertainties regarding the influences of CC in coastal aquifers were identified in Reference [11]. In Kish Island, Iran, the combined impacts of SLR with the associated land inundation and climate-induced variation of natural recharge to the aquifer system were analysed [12]. The results showed that the combined impact of SLR-induced land inundation and recharge rate variation is more significant compared to SLR impacts alone. A review of SLR impacts on SWI in coastal aquifers together with other factors, such as recharge rate variation, land inundation due to SLR, aquifer bed slope variation and changing landward boundary conditions [13], concluded that the impacts of these combined factors on SWI need to be further investigated. The NDA, like other coastal aquifers, is subjected to salinization threats.

Regarding NDA, [14] focused on a stretch between Ras El Bar and Gamasa along Egypt's northern coast. They found that the effect of SLR will be salinization of the NDA and they proposed artificial recharge through injection wells as a mitigation strategy. Further analysis of the impacts of SLR on SWI in the NDA was done by [15]. They concluded that large coastal areas will be totally submerged by SLR and that the shoreline will be shifted several kilometres inland. The research utilized a 2D FEFLOW model [16]. This modeling study used many assumptions due to a lack of available data. A 3D modeling approach was used by [17] to identify the effect of the increasing SLR, decreasing surface water level and increasing groundwater extraction on SWI in NDA. They found that under a combined scenario of increased groundwater extraction rates by 100%, an increased SLR by 100 cm and a decreased surface water level by 100 cm would cause the SWI to extend 79.5 km in the west and 92.75 km in the east from the shoreline by 2100.

While these studies have addressed some aspects of the SWI problem in the NDA, research covering the whole ND using 3D modeling to predict future scenarios of SWI has been quite limited. This approach is used in the current chapter to analyse the combined impact of SLR and excessive groundwater extraction in the NDA. Unlike previous studies that mainly presented the landside shift of the shore boundary and the dispersion zone due to SLR, the advantage of using a 3D model for this analysis is that it allows the determination of the full spatial distribution of SWI and the analysis of volumes of available fresh groundwater under different future scenarios. This enables a comparative analysis of the available volumes of different groundwater types (fresh water, light brackish, brackish and saline water) under different scenarios for the whole NDA. Moreover, for the first time, the volumes of each groundwater types are analysed per ND governorate (administrative regions in Egypt), according to their location in the ND. This may lead to more location-driven recommendations for mitigation and adaptation measures that could be implemented at the governorate level.

To carry out this analysis, a 3D variable-density groundwater flow model coupled with salt transport developed using the SEAWAT modeling system [18]. Because this is a regional model that is focused on freshwater resources in the NDA, it does not consider the dynamic coastline or the local coastal flooding processes. Thus, it is assumed that the horizontal position of the hydraulic head boundary at the coast is fixed. The model was calibrated by the groundwater salinity conditions in 2010, the year that serves as a baseline model and reference condition for future scenarios because most of the data required for a reliable analysis is sufficiently available for that year. This analysis proposes six scenarios to assess the impact of SLR versus groundwater extraction in the ND and determines which factor is causing increased groundwater salinization. The values assumed for the SLR and groundwater extraction rates are based on IPCC reports [19] and the Egyptian National Plan, as will be discussed later, in detail. In the first scenario (Sc.1), the model run until the year 2500 with no increase in SLR or groundwater extraction (same values as in 2010) to test the influence of time only on the complex groundwater system in the NDA (as SLR will not stop after 2100) and to investigate the natural autonomous salinization process [20]. The comparative analysis for the remaining five scenarios is carried out by running the model until the year 2100 with varying SLR and groundwater extraction rates. This analysis leads to a proposal of future groundwater extraction, levels for different ND governorates, which can be considered in future plans for the overall development of groundwater resources in the area.

4.3 STUDY AREA

The study area is the ND in the northern part of Egypt (Figure 4.1). It is the most fertile region in Egypt and is surrounded by highly arid desert. The ND begins approximately 20 km north of the Cairo governorate in the south and extends to the Mediterranean Sea in the north, covering an area of about 30 $\times 10^3$ km^2. In the west, it is bounded by the Alexandria governorate and in the east, by Port Said. The ND contains 11 governorates that have economic, cultural and agricultural importance to Egypt. The locations and the names of the ND governorates are presented in Figure 4.1.

The NDA is a large, semi-confined aquifer. It has Quaternary deposits that are classified into Holocene and Pleistocene strata [21]. The average thickness of the Holocene is 25 m. It reaches around 50 m close to the sea and vanishes towards the ND fringes in the south. The Pleistocene is the main aquifer of the ND [22]. Its thickness varies from 200 in the south to 1000 m in the north. It is composed of sand and gravel with occasional clay lenses and it is underlain by a Pliocene layer composed of marine limestone and shale [2, 3, 21].

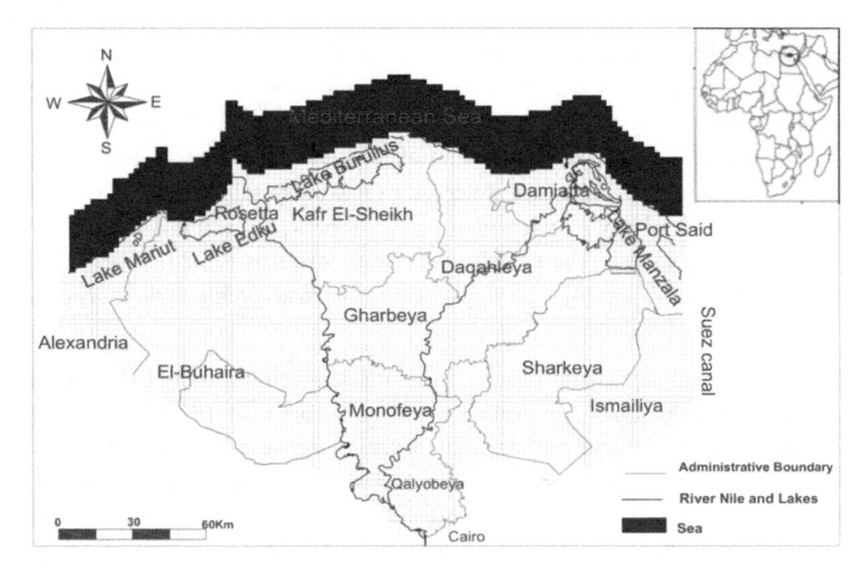

Figure 4.1. The study area of the ND

4.4 METHOD

4.4.1 Numerical model

A 3D variable-density groundwater flow and coupled salt transport model for the NDA was developed using SEAWAT [18]. The model captures the situation in the year 2010, since for this year, most data is available. It includes a large amount of different types of hydrogeological data collected from several sectors and research institutions within MWRI.

The model domain is discretized using 100 rows and 150 columns horizontally (grid cell sizes 2×2 km^2) and 21 modeling layers in the vertical direction, which enables sufficiently detailed simulations of salinity variations to be carried out under different conditions. The Mediterranean Sea is represented as a constant head and constant salinity boundary. The Suez Canal in the east is considered to be a no-flow boundary. The last (deepest) modeling layer of the 21 layers represents the impermeable Pliocene, which underlays the Pleistocene aquifer. This last modeling layer of the 21 layers also contains salinity sources arising from the dissolution of marine deposits present in the Pliocene [23]. The first (shallowest) modeling layer of the 21 layers represents the

Holocene layer which is specified as having a constant horizontal hydraulic conductivity of 0.25 m/day. For the next 19 modeling layers, the horizontal hydraulic conductivity values vary from 15 to 150 m/day. The anisotropy of hydraulic conductivity is considered to be 10% [24]. The effective porosity for the Holocene is specified as having a constant value of 40% and it varies from 12% to 28% in the Pleistocene. The main source of recharge in the NDA is the excess irrigation water at the agricultural zones. This spatially varying recharge, together with spatially varying water levels and salinity concentrations in the main irrigation canals, was specified to the model to ensure proper characterization of the groundwater entering the modeling domain. The majority of canals have water with a maximum salinity of 0.3 kg/m^3 but in some locations, the salinity concentration reaches 0.65 kg/m^3 and this is even greater towards the Mediterranean. The overall groundwater extraction in 2010 was estimated to be about 4.9x10^9 m^3/year [4]. The wells are distributed according to their depth in the corresponding modeling layer.

The model was calibrated with observed salinity data coming from 155 observation wells. The calibration results provided a Root Mean Square Error (RMSE) of 0.2 kg/m^3. The absolute difference and the standard deviation between the simulated and the observed concentrations of the salinity in the ND were calculated to be 0.14 kg/m^3 and 0.11 kg/m^3, respectively. These values are quite small, so the developed model is considered to be sufficiently reliable for future analyses. Further details about the model development and calibration results are available in [18].

4.4.2 Future scenarios

One of the most important factors that control the development scenarios is the population growth rate, as it represents the main condition that directly affects future economic and social progress. According to the Egyptian Central Authority for Public Mobilization and Statistics [25], the population in Egypt has raised by approximately from 52 million capita in 1990 to 84 million capita in 2010. The growth rate decreased from 2.75% per year in the period 1976–1986 to 2.02% per year during the period 1996–2006. The total population is predicted to reach 200 million capita in 2100 [26]. These population growth rates have led to corresponding increases in water needs for different water-consuming sectors. Moreover, in the future, the ascending gap between water demands and supply will result in more tension in negotiations over transboundary water projects on the Nile [27]. Groundwater extraction rates are likely to increase over time too, assuming the continuation of the almost linear trend of increase in groundwater extraction that follows population growth, as presented in Figure 4.2. This trend of increased groundwater extraction was constructed based on data collected from the past, starting in 1980 and then projecting until 2100 [4].

It is expected that CC will substantially raise the sea level. The Intergovernmental Panel on CC [19] emphasized the potential significance of SLR and predicted the global mean

SLR by the year 2100 to be in the range 0.26–0.82 m compared to the earlier foreseen 0.18–0.59 m [28]. It was reported in Reference [29] that melting processes in Antarctica could contribute to a SLR of 1.05 ± 0.3 m by 2100 under RCP 8.5. A probability density function of global SLR in 2100 was constructed in Reference [30]. It was found that the probability of having a SLR of more than 1.8 m is less than five percent, with a median probability of 0.8 m. Their findings were based on process models combined with experts' opinions. They also stated that other lines of evidence are needed to justify any higher estimates of SLR for 2100. Recently, the probability density functions for extreme scenarios of global SLR in 2100 based on extreme mass loss from ice sheets using numerical simulations with a process-based model have been provided [31]. The median of their probability distribution was 0.73 m. All these high SLRs will lead to severe impacts on SWI in coastal aquifers.

Future salinization of the groundwater resources in the NDA is a complex process and its prediction has a high degree of uncertainty. Consequently, different scenarios are proposed in this chapter to identify a wide spectrum of possible future adaptation strategies. These scenarios cover extreme projections to estimate future changes in the NDA in terms of the salinity distribution over the next hundred years as a function of SLR and groundwater extraction rates. Our estimated $12x10^9$ m^3/year of groundwater extraction in 2100 (Table 4.1) is consistent with the projections in the future scenarios of the MWRI national plan. Regarding the SLR, the selected ranges in the future scenarios (see again Table 4.1) were chosen by considering the SLR reported in Reference [19] and the median of the probability density functions for extreme scenarios of global SLR in 2100 reported in the literature review stated above.

Sc.1 (long run) is different from all of the other proposed scenarios. There is neither a SLR, nor an increase in groundwater extraction (the same values as in 2010 are used). This scenario analyses the impact of time only (until 2500), focusing on the autonomous salinization process, as it is known that the salinity distribution in coastal groundwater systems lags behind the current boundary conditions [20, 32, 33]. Thus, this scenario is not used for comparisons with the other scenarios. The remaining five future scenarios are combinations of different SLR and groundwater extraction rates.

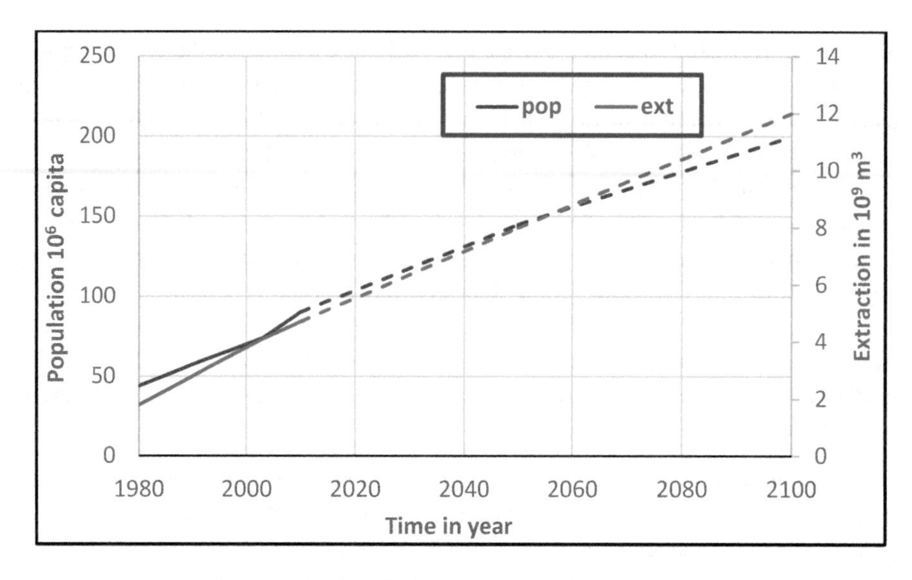

Figure 4.2. Population and groundwater extraction in the NDA as a function of time: current trend to 2010 (solid lines) and projections to 2100 (dashed lines) [26]

Table 4.1. Future scenarios for SLR and groundwater extraction

Scenario (Sc.)	SLR (m)	Extraction (10^9 m^3/year)	Time (year)
Reference	0	4.9	2010
Sc.1 Long run	0	4.9	2500
Sc.2 Extreme	1.5	12	2100
Sc.3 Moderate	1	8	2100
Sc.4 Restrictive	0	4.9	2100
Sc.5 High extraction	0	12	2100
Sc.6 High SLR	1.5	4.9	2100

Sc.2 (extreme) was selected to show the impact of the combined effect of a large SLR and extreme groundwater extraction on the NDA. The proposed population will reach 200 million per capita [26]. As the population growth continues with no birth control, a significant increase in groundwater extraction rate is expected 12×10^9 m^3/year to fulfil the high water demands (Figure 4.2). No control over groundwater extraction is assumed. In this scenario, the SLR is assumed to be 1.5 m, which is also considered extreme based on the analyses in the literature review presented earlier.

Sc.3 (moderate) takes birth control and a reduced population growth rate into account. It is associated with moderate control of groundwater extraction. The population is estimated to be about 146 million capita [26]. According to the national future plan of

MWRI in 2100, the government will have moderate investments in reclamation projects. The groundwater extraction will be $8x10^9$ m^3/year. The government will control the unplanned groundwater extraction and benefit from the solar desalination plants that are planned to be constructed. A moderate SLR of 1 m is also assumed.

Sc.4 (restrictive) is a very optimistic scenario as it assumes that no increase in groundwater extraction will occur, due to control by high financial penalties for any non-authorized groundwater extraction. In this scenario, the government depends on water resources other than groundwater. This scenario was selected to show the results of completely prohibiting groundwater extraction Also, it is assumed that there will be no SLR. In fact, this scenario has same conditions as Sc.1 (long run) but it is analysed until 2100.

In Sc.5 (high extraction), the government encourages investments and land reclamation which leads to a dramatic increase in groundwater extraction and hence, increases groundwater salinization. In this scenario, the groundwater extraction rates are proposed to reach again $12x10^9$ m^3/year by 2100. This scenario was selected to examine the impact of groundwater extraction alone, so the assumed SLR is 0 m.

On the contrary, the objective of Sc.6 (high SLR) is to measure the impact of SLR only. The groundwater extraction rate is same as now (2010) and a SLR of only 1.5 m is assumed. Scenarios 5 and 6 help to identify which is the main driving factor in the SWI process: SLR or groundwater extraction

All six scenarios consider that the Egyptian policy towards cooperation with the Nile basin countries will continue to be the same and consequently, water levels in canals and water use for agriculture will remain constant. The increase of groundwater extraction in different scenarios is without any spatial variation compared to 2010. This study analysed how the NDA will develop in the future with current spatial distribution of wells. Each existing well field will gradually increase the groundwater extraction until 2100. Further studies may investigate different spatial distributions of groundwater extraction. The initial conditions of the model were taken from the calibrated model's results for 2010. The model was run for 90 years for scenarios 2, 3, 4, 5 and 6 to predict the future SWI in 2100. For Sc.1, the model was run for 490 years to predict the SWI in the year 2500.

4.5 WATER SHARING ARRANGEMENTS

4.5.1 The whole Nile Delta

The total volume of groundwater in the NDA up to the hydrogeological base was estimated to be approximately $4050x10^9m^3$ covering an area of approximately $35x10^3$ km^2. This agrees with Reference [15] who estimated the groundwater volume of the

NDA to be around 3600x10^9 m^3 but with a smaller modeled area (24x10^3 km^2). According to [34], groundwater can be classified into different types with respect to the salinity level. Fresh water has a salinity of 0–1 kg/m^3, light brackish has 1–5 kg/m^3, brackish has 5–30 kg/m^3 and saline water has salinity greater than 30 kg/m^3. Figure 4.3 shows the different proportions of the groundwater types in 2010, together with their spatial distribution within the ND. Fresh water accounts for 37%, light brackish 10%, brackish 21% and saline water 32%. The volume of total available fresh water is approximately 1290x10^9 m^3 and the volume of light brackish water is about 421x10^9 m^3. Table 4.2 provides estimates of the groundwater volume in $\times 10^9$ m^3 occupied by the four groundwater types for the current conditions in 2010 and for the six analysed scenarios.

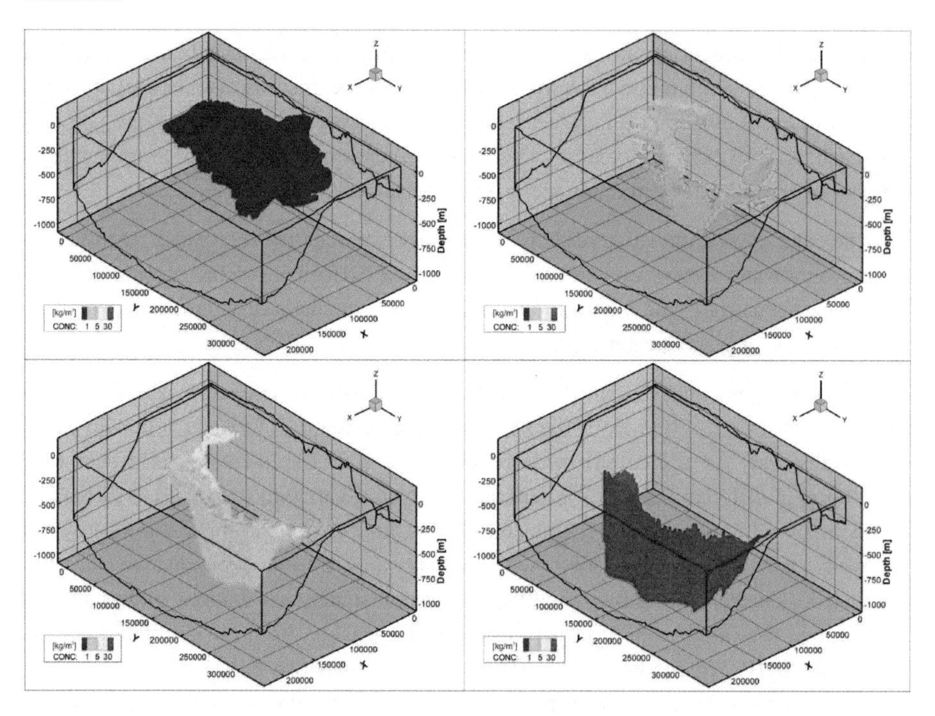

Figure 4.3. Spatial distribution of different groundwater types in the ND in 2010 (concentration in kg/m^3)

Table 4.2. Groundwater volume in $\times 10^9$ m^3 in the NDA for the four groundwater types

Gw. Types kg/m^3	Current 2010 Ref.	Sc.1 2500 Long Run	Sc.2 2100 Extreme	Sc.3 2100 Moderate	Sc.4 2100 Restrictive	Sc.5 2100 High Ext.	Sc.6 2100 High SLR
Fresh water 0–1	1290	893 −31%	1049 −18.7%	1119 −13.3%	1190 −7.7%	1090 −15.5%	1147 −11.1
Light brackish 1–5	421	436 +3.6%	434 +3%	432 +2.7%	431 +2.4%	433 +2.9%	432 +2.8%
Brackish 5–30	829	1051 +26.8%	900 +8.5%	888 +7.1%	886 +6.9%	894 +7.9%	890 +7.4%
Saline water >30	1513	1734 +14.6%	1691 +11.7%	1600 +5.7%	1548 +2.2%	1631 +7.7%	1611 +6.4%

It is clear from Table 4.2 that when only the influence of time is considered, the groundwater system becomes saltier. In Sc.1, the fresh groundwater volume decreases significantly by 31%, while the light brackish, brackish and saline groundwater volumes have corresponding increases. In this case, because of the long simulation period, the changes in salinization volumes are large in comparison with the other scenarios. The significance of the results from this scenario is that they demonstrate that even without any SLR or increased groundwater extraction, the salinization of the NDA will continue. It was shown by [18] that the NDA has not reached equilibrium yet and that this complex groundwater system is characterized by slow hydrogeological variations that bring significant impacts only after a long period of time [24]. For this reason, this scenario, in fact, raises a warning alarm. It shows the need for the planning of continuous adaptation strategies that prevent the accumulation of inland groundwater salinization. There are a number of adaptation strategies that could be applied in the ND [10], for example, artificial recharge, the extraction of saline and brackish groundwater and the modification of pumping practices through the reduction of withdrawal rates or adequate relocation of groundwater extraction wells.

For Scenarios 2 to 6, the models were run up until 2100 with different SLR rates and/or groundwater extraction rates. For Sc.2 (extreme), the fresh water decreased significantly by 18.7% in 90 years, while the saline volume increased by approximately 12%. These values are the highest among Scenarios 2 to 6, because Sc.2 shows the combined effect of an extreme SLR and groundwater extraction rate. Sc. 3 (moderate) shows modest values for fresh water volume loss (−13.3%) and saline groundwater gain (+5.7%).

Under very optimistic conditions, Sc.4 (Restrictive) has a small saline groundwater volume increase (+2.2%). The fresh groundwater loss (–7.7%) is the lowest among all other scenarios. To achieve this, however, very rigid control of groundwater extraction is necessary. Alternative, unconventional sources of water are required, possibly in combination with water saving and planting of crops that are more resistant to salt. Figure 4.4 shows the percent of change in different scenarios for the analysed groundwater types with respect to the 2010 baseline values.

In Sc.5 (high extraction), the fresh water volume decreases by about 15.5%. That decrease is second in severity after Sc.2 (extreme) with approximately 18.7%. This is also demonstrated in Figures 4.5 and 4.6 below, indicating that the largest part of the decrease in volume comes from groundwater extraction, not SLR.

Table 4.2 shows a decline in fresh water in Sc.6 (high SLR) to about 11%. This value is less than that in Sc.5 discussed above, meaning that SLR has a smaller effect on the groundwater volumes in the whole NDA compared to groundwater extraction.

Figure 4.5 shows different cross-sections of the NDA for Scenarios 2, 5 and 6. It is clear that the salinity front advances the most in Sc.2 (extreme) followed by Sc.5 (high extraction) and Sc.6 SLR. The scenarios have the same ordering with respect to the size of the dispersion zone.

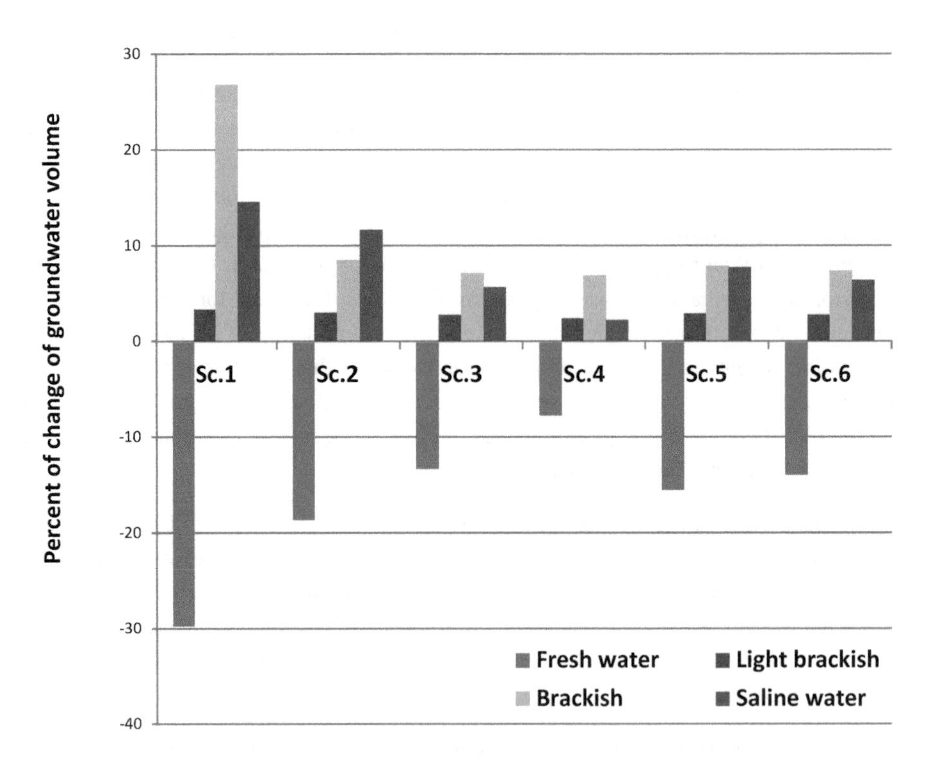

Figure 4.4. The percent of change of groundwater types in different scenarios, with respect to the 2010 baseline values

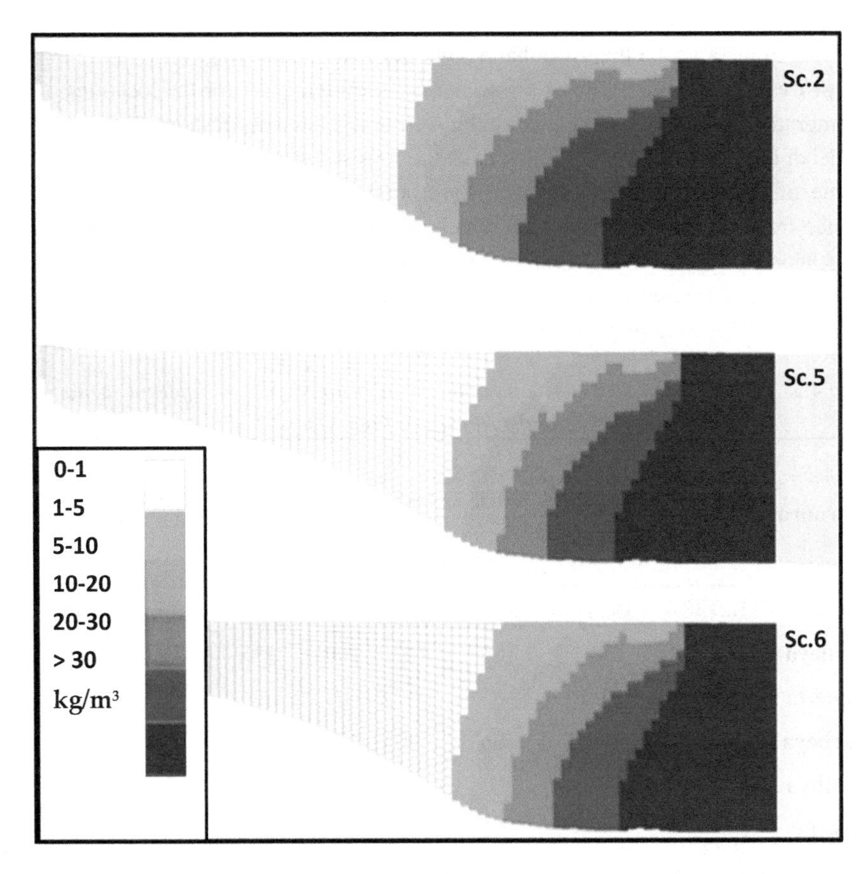

Figure 4.5. Cross-sections in the middle of the ND with different salinity distributions in different scenarios in the year 2100

4.5.2 The Nile Delta governorates

As stated in the study area section, the ND management system is organized into a number of governorates. The Egyptian governorates are administrative divisions. They are the second tier of the country's jurisdiction hierarchy, below the national government. Each governorate is administered by a governor, who is appointed by the President of Egypt [35]. The groundwater sector in Egypt is managed jointly by the MWRI and the governor in each governorate. Consequently, it is necessary to study salinization patterns with respect to each governorate, because each governorate has different hydrogeological parameters and more importantly, different agricultural and water resource management activities (groundwater extraction and irrigation) and thus, different groundwater type and salinization patterns.

Some of these governorates are coastal (e.g., Alexandria, Damietta), while others are far away from the zone where brackish and saline groundwater are present.

It is clear from Table 4.3 that El Buhaira, Gharbeya and Sharkeya governorates have the highest freshwater volumes in all six scenarios (see Figure 4.1 for the locations of the governorates). On the other hand, coastal governorates have no fresh groundwater. Kafr El Sheikh and Ismailiya governorates have the lowest fresh groundwater volumes. The volume of fresh groundwater in each governorate is dependent on its location and distance from the coast, as well as the thickness of the aquifer, the governorate's area, the reclamation projects and the number of groundwater extraction wells. Here, again, it is clear that Sc.2 (extreme) has the highest impact of the governorates on salinization in 2100, followed by Sc.5 (high extraction) and Sc.6 (high SLR).

Table 4.3. The fresh groundwater volume in 10^9 m^3 with respect to different scenarios in all governorates of the ND

Governorate	Area 10^3 km^2	Ext. 2010 10^6 m^3/year	2010	Sc.1	Sc.2	Sc.3	Sc.4	Sc.5	Sc.6
El-Buhaira	10.130	1931	258	180	228	253	250	234	243
Daghleya	3.5	114	160	62	100	135	143	118	127
Damietta	1.029	362.5	0	0	0	0	0	0	0
Gharbeya	1.942	291.7	306	200	274	289	294	279	286
Ismailiya	2.10	163.8	31	8	16	26	28	20	24
Kafr El Sheikh	3.437	0.9	24	0	11	19	20	16	18
Monofeya	2.543	791.5	120	105	113	117	118	114	116
Qalyobeya	1.124	408.2	66	51	54	61	63	56	60
Sharkeya	4.18	681.5	318	210	276	295	302	281	290
Alexandria	2.679	2.4	22	5	7	14	15	9	13
Portsaid	1.351	154.2	57	96	45	50	51	46	49

Figure 4.6 shows that for all scenarios, the fresh groundwater decreases at a lower rate in the southern governorates, like Qalyobeya and Monofeya, compared to the northern governorates, like Kafr El Sheikh. This is mainly because these governorates are far from the coastal zone. Meanwhile, the volume of fresh groundwater decreases at a high rate in the governorates Sharkeya and El Buhaira, which suffer from the combined effect of excessive groundwater extraction and SLR. The analysis of the spatial distribution of salinity indicates that in these governorates, the northern parts are already in a critical condition, while some continued groundwater extraction can still be allowed in the southern regions. As we can see from Figure 4.6, Ismailiya governorate shows a

significant drop of the (relatively small) fresh groundwater volume in all different scenarios. This governorate is already stressed by severe groundwater extractions, so future scenarios further intensify its criticality regarding available fresh water resources. In Dagahleya, the drop is also significant, although the absolute volume of fresh groundwater is larger. Coastal governorates (Kafr El Sheikh, Damietta and Alexandria) are very vulnerable to SLR. Damietta already has no fresh groundwater, while Kafr El Sheikh shows a severe reduction of fresh groundwater in the future in all scenarios.

Figure 4.7 shows the distribution of different groundwater types for selected governorates in the current situation (2010) and in 2100 under Sc.2 (Extreme). The Monofeya governorate has no saline or brackish groundwater, while the coastal governorates, for example, Kafr El Sheikh and Damietta, contain mainly brackish and saline groundwater.

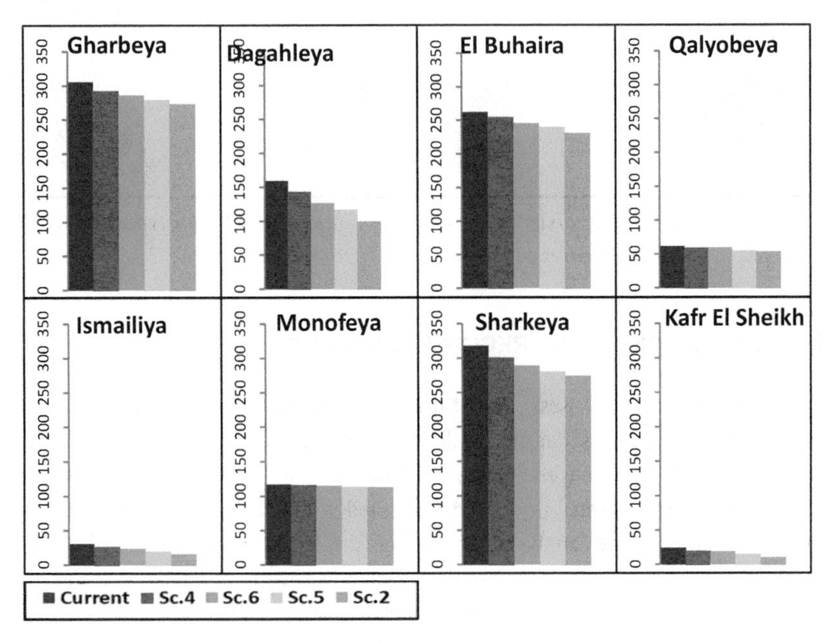

Figure 4.6. Block diagrams representing the fresh groundwater volume for different scenarios in selected governorates

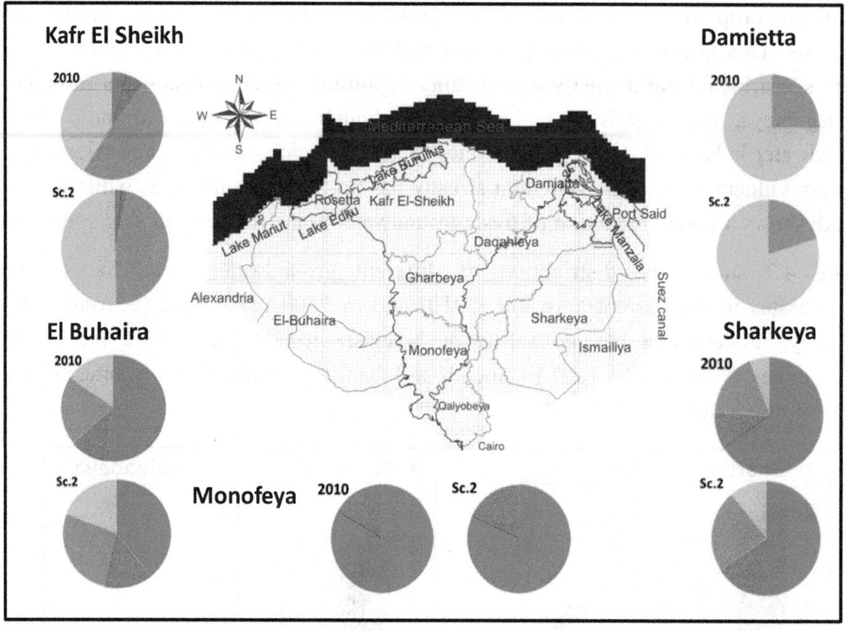

Figure 4.7. Proportional distribution of the different groundwater types for selected governorates in the current situation and under Sc.2

4.6 CONCLUSIONS AND RECOMMENDATIONS

This chapter presented an assessment of the impacts of SLR and groundwater extraction on SWI in the NDA. Six scenarios of different combinations of SLR and groundwater extraction from the NDA were proposed. The analysis was carried out using a 3D variable-density groundwater flow model coupled with salt transport. The groundwater salinity conditions were analysed by using the change in volume for the four groundwater types in the whole ND and for each ND governorate separately. The results clearly demonstrate that the effect of groundwater extraction is more severe than the SLR. However, SLR is linked to CC, which is beyond the direct control of the Egyptian government and imposes an extra burden on the groundwater system of the NDA, especially when it is combined with excessive groundwater extraction. Salinization will intensify in the future if adaptation and mitigation measures are not implemented.

It is recommended that groundwater extraction is banned in northern coastal governorates that are very sensitive to SLR and groundwater extraction. No groundwater extraction or investment in agriculture should be made in the coastal governorates Ismailiya and Dagahleya due to their shortage in fresh groundwater. In the southern governorates, like Qalyobeya and Monofeya, the groundwater extraction of fresh water can continue, as the results indicate that these governorates have only fresh and light brackish water and do not have any brackish or saline water in all future scenarios. Groundwater extraction can be allowed only in the southern parts of the Sharkeya and El Buhaira governorates, because their northern parts are already in a critical condition. Table 4.4 summarizes the recommended locations for future groundwater extraction in different governorates of the ND.

Table 4.4. A Summary of the recommended groundwater extraction locations in the ND governorates

Governorate	Recommendation	Reason
-Coastal governorates Kafr El Sheikh Alexandria Damietta Port Said	Ban groundwater extraction Search for alternatives	-Limited or no fresh groundwater -Sensitive to SWI
-Middle governorates El Buhaira Gharbeya Sharkeya	Extraction in the southern region	-Fresh water is decreasing rapidly -Suffer from combined effects of excessive groundwater extraction and SLR -Have the highest fresh water volumes
Ismailiya Dagahleya	No groundwater extraction No investment in agriculture that relies on groundwater	-A huge drop of fresh water volume between different scenarios -SWI and extraction that intensify the criticality of its small fresh water volume
-Southern governorates Qalyobeya Monofeya	Groundwater extraction is recommended	-Fresh water is decreasing slowly -Far from SWI

For strategic planning of groundwater management in Egypt until 2100, it is recommended that future groundwater extraction distribution is adjusted. The vulnerability of governorates in terms of the available fresh groundwater volume (under stress) should be taken into consideration when designing future water management adaptation plans. Taking serious measures, such as closing illegal wells, implementing full control of new authorization permits for groundwater extraction in the NDA and giving strict fines for any violation of groundwater extraction regulations, will be required to reduce the salinization problem. Additionally, future innovations and technologies, such as wastewater reuse, crop management and desalinization, should be actively pursued. Moreover, immediate initialization of development projects that could protect groundwater salinization is needed, for instance, desalination and rain harvesting projects in the northern coastal region.

Finally, it should be recognized that SWI is a long-term process which may led to the future deterioration of fresh groundwater resources in the NDA, even beyond 2100. Sc.1 which runs until 2500 indicates that time alone can bring further SWI into the NDA. If uncontrolled development of groundwater resources continues while control and adaptation measures are not taken into account, this valuable fresh groundwater resource will be impaired to an extent that negatively affects the overall socioeconomic development of the country.

References

1 EGSA. *Egyptian General Survey and Mining: Topographical Map cover ND , Scale 1: 2 000 000*; Egyptian General Survey and Mining (publishing center): Cairo, Egypt, 1997.

2 SADS2030. *Sustainable Agricultural Development Strategy, Ministry of Agriculture and Land Reclamation of Egypt*, 1st ed., 197, 2009.

3 Farid, M.S.M. Nile Delta groundwater study. Msc. thesis, Cairo Uni., Cairo, Egypt, 1980.

4 Mabrouk, M.B.; Jonoski, A.; Solomatine, D.; Uhlenbrook, S. A review of seawater intrusion in the ND groundwater system—The basis for assessing impacts due to CC s, SLR and water resources development. *Nile Water Sci. Eng. J. 10*, 46–61, ISSN 2090-0953, 2017.

5 Diab, M.S.; Dahab, K.; El Fakharany, M. Impacts of the paleohydrological conditions on the groundwater quality in the northern part of ND, The geological society of Egypt. *Geol. J. 4112B*, 779–795, 1997.

6 Custodio, E. Aquifer overexploitation: What does it mean? *Hydrogeol. J.* 2002, *10*, 254–277, doi: 10.1007/s10040-002-0188-6.

7 Yihdego, Y.; Panda, S. Studies on nature and properties of salinity across globe with a view to its management-a review. *Glob. J. Hum.-Soc. Sci. Res. 17*, 31–37, 1, ISSN 0975-587X, 2017.

8 Oude Essink, G.H.P.; Van Baaren, E.S.; De Louw, P.G.B. Effects of CC on coastal groundwater systems, a modeling study in the Netherlands. *Water Resour. Res. J. 46*, W00F04, doi: 10.1029/2009WR008719, 2010.

9 Yihdego, Y.; Al-Weshah, R.A. Assessment and prediction of saline seawater transport in groundwater using 3-D numerical modeling. *Environ. Processes J. 4*, 49–73, doi: 10.1007/s40710-016-0198-3, 2016.

10 Gorelick, S.M.; Zheng, C. Global change and the groundwater management challenge. *Water Resour. Res. J. 51*, 3031–3051, doi: 10.1002/2014WR016825, 2015.

11 Ojha, L.; Wilhelm, M.B.; Murchie, S.L.; McEwen, A.S.; Wray, J.J.; Hanley, J.; Massé, M.; Chojnacki, M. Spectral evidence for hydrated salts in recurring slope linear on Mars. *Nat. Geosci. J. 8*, 829 doi: 10.1038/ngeo2546, 2015.

12 Mahmoodzadeh, D.; Ketabchi, H.; Ataie-Ashtiani, B.; Simmons, C.T. Conceptualization of a fresh groundwater lens influenced by CC, A modeling study of an arid-region island in the Persian Gulf, Iran. *Hydrol. J. 519*, 399–413, doi:10.1016/j.jhydrol.2014.07.010, 2014.

13 Ketabchi, H.; Mahmoodzadeh, D.; Ataie-Ashtiani, B.; Simmons, C.T. Sea-level rise impacts on seawater intrusion in coastal aquifers: Review and integration. *Hydrol. J.* doi:10.1016/j.jhydrol.2016.01.083, 2016.

14 Nofal, E.R.; Fekry, A.F.; El-Didy, S.M. Adaptation to the impact of SLR in the ND coastal zone, Egypt. *Am. Sci. J.* 10, 17–29, ISSN: 1545-1003, 2014.

15 Sefelnasr, A.; Sheriff, M.M. Impacts of seawater rise on seawater intrusion in the NDA, Egypt. *Groundw. J.* 52, 264–276, 2014.

16 Sherif, M.M.; Sefelnasr, A.; Javadi, A. Incorporating the concept of equivalent fresh water head in successive horizontal simulations of seawater intrusion in the NDA, Egypt. *Hydrol. J.* 464–465, 186–198 doi:10.1016/j.jhydrol.2012.07.007, 2012.

17 Abdelaty, I.M.; Abd-Elhamid, H.F.; Fahmy, M.F.; Abdelaal, G.M. Investigation of some potential parameters and its impacts on SWI in NDA. *Eng. Sci. J.* 2014, 42, 931–955, doi: 10.1038/nature17145, 2014.

18 Mabrouk, M.; Jonoski, A.; Oude Essink, G.H.P.; Uhlenbrook, S. Assessing the Fresh–Saline Groundwater Distribution in the Nile Delta Aquifer Using a 3D Variable-Density Groundwater Flow Model. *Water J.*, 11, 1946, 2019.

19 Oppenheimer, M., B.C. Glavovic , J. Hinkel, R. van de Wal, A.K. Magnan, A. Abd-Elgawad, R. Cai, M. Cifuentes-Jara, R.M. DeConto, T. Ghosh, J. Hay, F. Isla, B. Marzeion, B. Meyssignac, and Z. Sebesvari, 2019: Sea level rise and implications for low-lying islands, coasts and communities. In: IPCC special report on the ocean and cryosphere in a changing climate, 2019.

20 Delsman, J.R.; Hu-a-ng, K.R.M.; Vos, P.C.C.; De Louw, P.G.B.; Oude Essink, G.H.P.; Stuyfzand, P.J.; Bierkens, M.F.P. Paleo-modeling of coastal SWI during the Holocene, an application to the Netherlands. *Hydrol. Earth Syst. Sci. J.* 18, 3891–3905. doi:10.5194/hess-18-3891-2014.

21 Sestini, G. ND: A review of depositional environments and geological history. *Geol. Soc. Lond. Spec. Publ.*, 41, 99–127, doi:10.1144/GSL.SP.041.01.09,1989.

22 Morsy, W.S. Environmental management to groundwater resources for ND region. Ph.D. Thesis, Faculty of Engineering, Cairo University, Cairo, Egypt, 2009.

23 Van Engelen, J.; Oude Essink, G.H.P.; Kooi, H.; Bierkens, M.F.P. On the origins of hypersaline groundwater in the NDA. *Hydrol. J.* 560, 301–317. doi:2018.03.029/hydrol10.1016.

24 Bear, J. Hydraulics of Groundwater; McGraw-Hill Book Company, New York, United States of America, 1979, p. 592, ISBN 0-486-45355,2018.

25 CAPMAS. The Central Authority for Public Mobilization and Statistics, Egypt, Egypt in Numbers; Ministry of Communication and Information Technology (publishing center): Cairo, Egypt, March 2010.

26 United Nations, Department of Economic and Social Affairs, population division. World population prospects: The 2015 revision, key findings and advance tables, United Nations, New York , working paper no. ESA/P/WP.241. Available online: https:// esa. un. org/ unpd/ wpp/ publications/files/key_findings_wpp_2015.pdf, 2015.

27 Yihdego, Y.; Khalil, A.; Salem, H.S. Nile Rivers Basin dispute: perspectives of the grand Ethiopian Renaissance Dam (GERD). Hum.-Soc. *Sci. Res. J.* 17, 1–21, ISSN2249-460, 2017.

28 Solomon, S.; Qin, D.; Manning, M.; Chen, Z.; Marquis, M.; Averyt, K.B.; Tignor, M.; Miller, H.L. Contribution of working group I to the fourth assessment report of the intergovernmental panel on climate change ; Cambridge Uni. press, Cambridge, United Kingdom and New York, United States of America, 2007.

29 DeConto, R.M.; Pollard, D. Contribution of Antarctica to past and future sea level rise. *Nature J.* 531, 591–597, doi: 10.1038/nature17145, 2016.

30 Jevrejeva, S.; Grinsted, A.; Moore, J.C. Upper limit for sea level projections by 2100. *Environ. Res. Lett. J.* 9, 104008, doi:10.1088/1748-9326/9/10/104008, 2014.

31 Le Bars, D.; Drijfhout, S.; de Vries, H. A high-end SLR probabilistic projection including rapid Antarctic ice sheet mass loss. *Environ. Res. Lett.J.* 12, 044013, doi:10.1088/1748–9326/aa6512, 2017.

32 Meisler, H.; Leahy, P.P.; Knobel, L.L. Effect of Eustatic sea level changes on saltwater-fresh water in the northern Atlantic Coastal Plain; US Government Printing Office, USGS *Water Supply Paper:* Alexandra VA, United States of America, 2255, 34,1984.

33 Larsen, F.; Tran, L.T.L.V.; Van Hoang, H.; Tran, L.T.L.V.; Christiansen, A.V.; Pham, N.Q. Groundwater salinity influenced by Holocene seawater trapped in incised valleys in the Red River delta plain. Nat. *Geosci. J.* 10, 376–381. doi:10.1038/ngeo2938, 2017.

34 Lide, D.R. (ed.) CRC Handbook of chemistry and physics, 86th ed.; CRC Press: Boca Raton, FL, USA, 2015; 8–71, 8–116. ISBN: 0-8493-0486-5, 2015.

35 Metz, H.C. (ed.) Egypt, A Country Study; retrieved 21 October 2016; GPO for the Library of Congress: Washington, DC, USA, 1990.

5 ADAPTATION MEASURES TO MITIGATE GROUNDWATER SALINIZATION THREATS IN THE NILE DELTA AQUIFER

The content of this chapter is a full reproduction of the submitted article:

Mabrouk, M.; Jonoski, A; Oude Essink, G.H.P.; Uhlenbrook, S. Adaptation measures to mitigate groundwater salinization threats in the Nile Delta Aquifer. *Water J.* 2020 (In review).

5.1 ABSTRACT

Nile Delta (ND) region is highly vulnerable to climate change (CC) impacts. Potential impacts on groundwater salinization may be very significant, because of the associated sea level rise (SLR). In combination with increased groundwater extraction, this may put large pressures on the fresh groundwater resources available in the Nile Delta Aquifer (NDA). This calls for an immediate planning of potential measures for dealing with these negative impacts.

The main objective of this research is to analyse different adaptive measures to groundwater salinization using a groundwater salinity model developed with the SEAWAT code. The model has been developed for the whole NDA, but the analysis is carried out for the Sharkeya governorate only.The research examines three adaptive measures techniques to retard or prevent the impact of saltwater intrusion (SWI) in the ND: well injection, brackish water extraction and changing of cropping patterns and irrigation practices. The main criterion used for comparison between the different adaptation measures is in terms of the overall freshwater availability Changing of cropping patterns and irrigation practices to more water-saving options seems to be very promising measure from a freshwater quantity point of view. However, the complexity of the problem requires likely implementation of a combination of such measures, with further analyses of appropriate locations for particular measures and agro-economic considerations.

5.2 INTRODUCTION

CC is expected to continue to take place over the next century and to exacerbate existing environmental problems in many countries [1]. In particular, coastal aquifers all over the world are expected to suffer from the impacts of SLR as a consequence of CC. However, it is important to recognize that impacts of SLR must be considered in relation to impacts of development, e.g. excessive extraction of groundwater [2]. Most coastal aquifers will be affected with groundwater salinization due to these combined effects and the NDA in Egypt is no exception. A number of economically important governorates in the ND, where most fertile land exits, will be exposed to these adverse impacts [3].

Excessive extraction started in the NDA during the 1980-ies, while the illegal and unplanned extraction increased intensively in the 1990-ies [4] stressing the scarcity and salinization of this valuable resource. These persisting problems need a proper adaptation management plan that could control and protect the NDA. That plan should be well designed to sustain that valuable resource against salinization threats, before it becomes too late to take adaptation measures against future expected negative impacts [2]. The sooner the adaptation activities begin, the lower the likely impacts.

Research on different CC adaptation measures for groundwater salinization has been published worldwide. [5] used analytical solutions to assess the interface between freshwater-seawater resulting from subsurface barrier partially embedded in a coastal aquifer. It was found that subsurface barrier can be only applied in areas with limited hydraulic gradients. [6] studied the impact of subsurface dams and found that they prevent SWI and reduced the salinity already existing due to SWI in the coastal aquifer. Well injection has attracted many researchers as it is considered a suitable adaptation measure to reduce SWI. [7] studied the effectiveness of recharging wells. Despite the negative drawbacks of this method related to the reduction in permeability in the injected area due to well clogging, it has been practiced in different countries around the world [8]. In 1985, [9] studied the freshwater-saltwater interface movement using multiple recharging wells. However, his solutions cannot be applied at the toe or within the saltwater wedge but only on the landward side of the interface. It was discovered that at those positions the effect of recharging wells is minimal for reducing groundwater salinization. [10, 11] used a series of injection wells on the freshwater-saltwater interface using a sharp interface finite element model. This study proved that a range of 60-90 percent of the prevention of SWI could be achieved depending on the number and spacing between the injection wells. [12] used SUTRA for simulation of SWI. They proved that injection wells method is a very effective adaptation measure. Their simulations results proved that it reduces the extent of SWI and seepage velocities. They declared that the system is closer to steady state conditions rather than the no remedial action by using injection wells.[13] recommends land reclamation, thus creating a foreland where a freshwater body may develop causing a delay to inflow of saline groundwater.

In Egypt, [14] studied the socioeconomic impact of SLR in Alexandria and Port Said governorates to assess the need and importance of adaptation. They concluded that the NDA region is highly vulnerable to SWI. They highlighted that the impact on agricultural resources is significant and the losses of cultural heritage cannot be estimated and calls for immediate serious adaptation plan. [15] designed an adaptation measure for the agricultural sector in the western region of the ND. Their study concluded that traditional knowledge compatible with local development requirements could be the gateway for using simple and low cost adaptation measures to meet local conditions. They found that making inexpensive adjustment such as increasing the water use efficiency of an agricultural system, appropriate soil drainage, land management and crop alternatives use will lessen the potential negative salinization impact on agriculture. [16] presented different mitigation methods for a strip of 15 km in the northern coast of the ND. These mitigations include artificial recharge through injection wells, impervious barriers, constrains on groundwater extraction and implementation of local monitoring network. They found that using injection wells will enhance the salinity in the injection zone and that regarding the impervious barrier, it will not be effective in long term simulations.

There is very limited research in the NDA dealing with adaptation to the increased salinization threats using numerical analysis. The research on adaptation in Egypt included only general recommendations without giving any concrete methodology. This calls the need for analysis that can provide adaptation measures to prevent or reduce groundwater salinization. Such analysis should be based on a reliable numerical model that can simulate most likely future salinization scenarios. The model will allow testing of different adaptation measures before salinization risks due to SLR and future extraction become unmanageable.

In this research, we examine three different adaptation measures: well injection, extraction and treatment of brackish water and changing of cropping patterns and irrigation practices To carry out this analysis, a 3D variable-density groundwater flow model was developed using the SEAWAT modeling system for the NDA [2]. The model was developed to represent the groundwater salinity conditions in 2010. For this year, most of the data required for reliable analysis were sufficiently available and a baseline model was developed to represent the salinity distribution in the NDA as a reference condition. The model was then used to simulate different future scenarios (till year 2100) consisting of combinations of SLR and increased groundwater extraction [17]. . The developed model was carried out for the whole NDA, but here, the same model is used to test the three proposed adaptation measures in one selected governorate – Sharkeya. It is characterized with high agricultural activities and large range of salinity concentration values in the aquifer. The results from the proposed adaptation measures are compared in terms of fresh groundwater availability.

5.3 STUDY AREA

The study area is the Sharkeya governorate in the eastern ND region (Figure 5.1). It has a population of about 6 million capita in an area of 4.18×10^3 km^2 [18]. It is considered as the third governorate in the net return from crop production in Egypt; it has the highest cultivated area in the ND and the highest water consumption rates [19]. The groundwater extraction rate is 681.5 Mm^3/year (the second highest governorate in the ND).

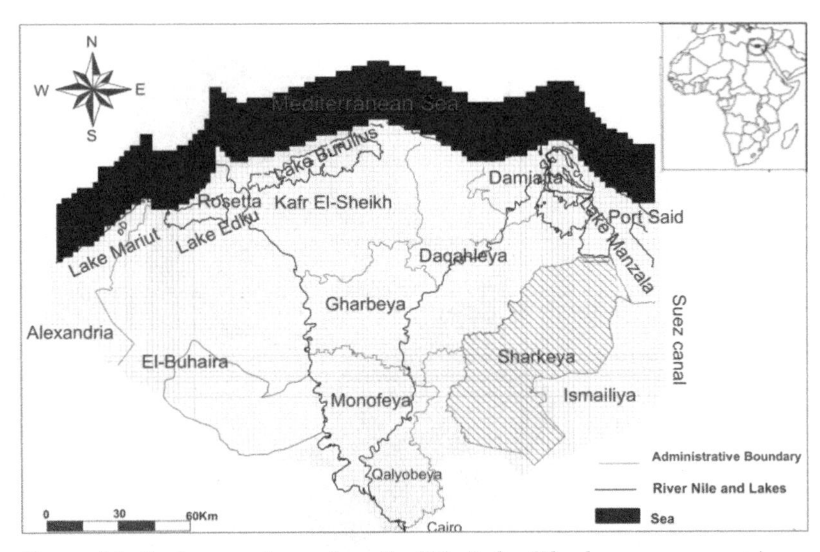

Figure 5.1. Study area chosen from the Nile Delta (Sharkeya governorate)

The Sharkeya governorate suffers from combined effects of excessive extraction and SLR due to its location [2]. Figure 5.2 shows the salinity distribution in the last layer of the aquifer. It is important as it shows that the Sharkeya governorate has a wide range of salinity concentration distribution that ranges from 0.2 to 21 kg/m^3. The northern part is rarely cultivated because the groundwater has high salinity. As shown from Figure 5.2, the extraction increases in 10th of Ramadan and El Salhia cities.

We chose the Sharkeya governorate for several reasons: the large amount of different types of hydro (geo) logical data to support our research of the groundwater system; the economic agricultural value of that governorate and the critical salinization distribution that reflects the urgency to adapt to (possibly even increasing) salinization risks [2].

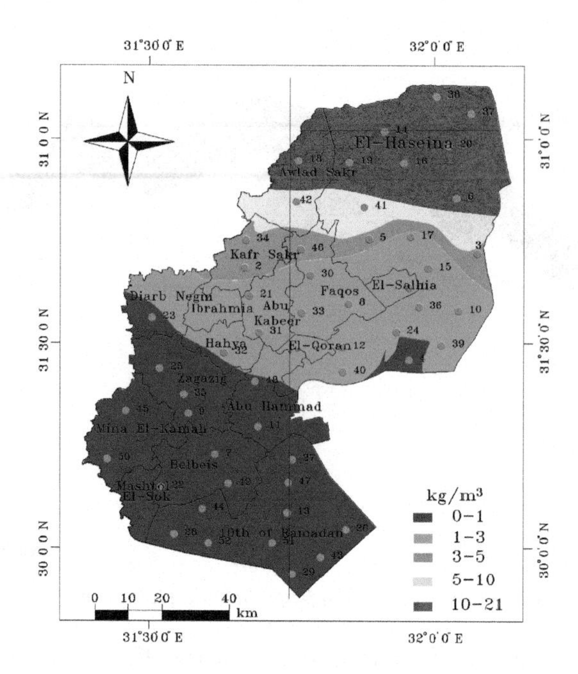

Figure 5.2. The salinity distribution in the last layer of the aquifer; the numbers indicate the observation wells

5.4 METHODOLOGY

5.4.1 Numerical model

A 3D variable-density groundwater flow was developed for the NDA [2]. The model was developed using large amount of different hydrogeological data gathered from various sectors and research institutions within MWRI, and from the private sector. It represents the salinity distribution in the NDA for year 2010.

In the vertical direction, the modelling domain was represented with 21 computational layers for simulation of the flow and salt transport processes in vertical and horizontal direction to show clearly the SWI and salinity variations under different conditions. Each model layer was discretized with 100 rows and 150 columns. The total area of the active cells covered by the model is about 35,000 km^2. The grid element size is approximately 2 km by 2 km covering the whole modeled area.

The top layer represents the Holocene clay layer. This layer is followed by 19 computational layers that represent the Pleistocene aquifer. The last computational layer represents the Pliocene Formation that underlays the Pleistocene aquifer. This last computational layer is needed to represent the dissolution impact of some salts and minerals from the underlying Pliocene Formation [20].

The horizontal hydraulic conductivity of the first modeling layer is specified as 0.25 m/day. For the next 19 modeling layers, the horizontal hydraulic conductivity values vary from 15 to 150 m/day. The anisotropy of hydraulic conductivity is considered to be 10% [8]. The effective porosity varies from 12% to 28% in the Pleistocene while it is specified as 40% in the Holocene layer. The main source of recharge in the NDA is the excess irrigation water as the rain water is almost neglected. The recharge rate ranges from 0.25 to 1.8 mm/day. This spatially varying recharge, water levels and salinity concentrations in the main irrigation canals, was specified to the model to ensure proper characterization of the groundwater entering the modeling domain.

The majority of the canals have a maximum salinity of 0.3 kg/m^3 but in some locations, the salinity concentration reaches 0.65 kg/m^3 towards the Mediterranean Sea. The RMSE is 0.2 kg/m^3. The absolute difference and the standard deviation between the simulated and the observed concentration of the salinity in the ND were calculated to be 0.14 kg/m^3 and 0.11 kg/m^3, respectively. These values are quite small, so the developed model is considered to be sufficiently reliable for future analyses. For further details on the model development and its verification, see [2]. Since the ND model used contains a rich data set for all hydraulic parameters, it is suitable for more detailed analysis on different ND governorates.

5.4.2 Adaptation method proposed

In this research, the SEAWAT code was used to study groundwater flow and different groundwater salinity concentrations in the Sharkeya governorate. Three adaptation measures are suggested to delay or prevent the impact of SWI in the NDA: well injection, brackish extraction and changing of cropping patterns and irrigation practices.

The adaptation measures proposed are based on the most likely moderate future scenario. This scenario assumed an increase in the extraction rate to reach 8 x10^9 m^3/year and SLR of 1 m. The values assumed for the SLR and groundwater extraction rates are based on [1] and the Egyptian National Plan, respectively. With the most likely future moderate scenario considered the current available fresh water volume in the Sharkeya governorate of 322x10^9 m^3 in 2010 will be decreased to 295x10^9 m^3 in 2100, which indicates a loss of fresh groundwater volume in the aquifer of about 27x10^9 m^3 after 90 years. The three adaptation measures are then assumed to be implemented during the same period and the lost /gained fresh groundwater volumes are reported.

The main points of comparison between the different adaptation measures are in terms of the available amount of freshwater in the aquifer, gain/loss in available freshwater, and additional freshwater outside the aquifer (which may be obtained with some of the adaptation measures). Salinity concentration (before and after adaptation measure) applied in the model is obtained, from which the fresh groundwater availability is assessed (with salinity concentration 0–1 kg/m^3). Advantages and disadvantages of each method are then discussed for initial comparison of the proposed measures. The economic feasibility of each adaptation measure should be further investigated in light of the benefits gained in reducing the SWI.

The required water for the proposed adaptation methods could be available through two methods:

1 Well injection

Well injection is a very popular adaptation measure that could mitigate or minimize SWI in coastal aquifers [13, 21, 22]. Use of tertiary treated water as a source for artificial recharge has been practiced in several countries, such as the USA, Australia, the Netherlands and other European countries [23, 24, 25, 26]. It is increasingly being considered in other countries facing water scarcity, such as those from the Middle East and the MENA region [27, 28]. The main target of proposing this adaptation measure is to create a hydraulic barrier by raising the piezometric head of the NDA, which will prevent the SWI. In this research, we carried out numerical simulations to determine the effect of application of this measure.

One of the important considerations when using injection wells for aquifer recharge are wells' number and spatial distribution, because they influence the effectiveness of the measure. Most effective results are achieved when the injection takes place in the fresh groundwater body, close to the interface with brackish groundwater, but many different injection wells configurations still need to be tested on actual implementation sites. This is a problem that can be addressed by coupled simulation-optimization approach, in which groundwater models are coupled with optimization algorithms for determining optimal management strategies [29].

In this analysis, different schemes of injection rates and distribution schemes of the wells were tested and evaluated with the model. After several trials, the most promising scheme for salinity concentration adaptation and hydraulic head was by proposing of 66 wells at a distance of 2 km between the wells. The depth of the wells is about 300 m. The injection rate is 2000 m^3/day for a period of 90 years. The total injected recharge through the whole scheme is around $50x10^6$ m^3/year which represents about 60% of the available tertiary treated wastewater (Figure 5.4).

The amount of the overall collected sewage in the Sharkeya governorate is estimated at $148x10^6$ m^3/year. The treated tertiary sewage water is $81x10^6$ m^3/year [16]. There are 29 plants for production of treated sewage water that are distributed all over the governorate (Figure, 5.3). The overall cost for the production of 1 m^3 of treated tertiary sewage water is around 2-3 EGP (equal approximately to 10-15 cent in 2020) [18]. The treated sewage water cannot be used directly in agriculture due to the environmental regulations and laws of the Ministry of Environment in Egypt [Law no 4, 1994] and the parliament decision [Decree no 603, 2002]. Consequently, huge amount of treated water could be reused and injected by wells into the NDA.

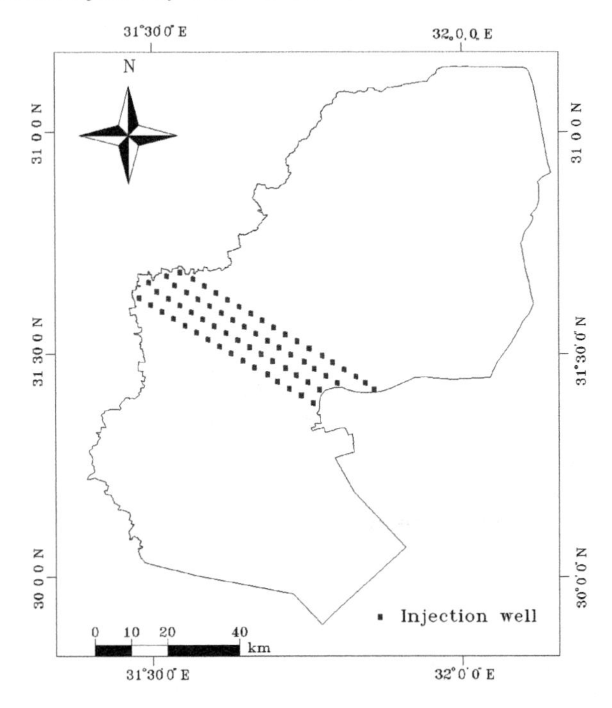

Figure 5.3. The distribution scheme of the injection wells in the Sharkeya govenerate

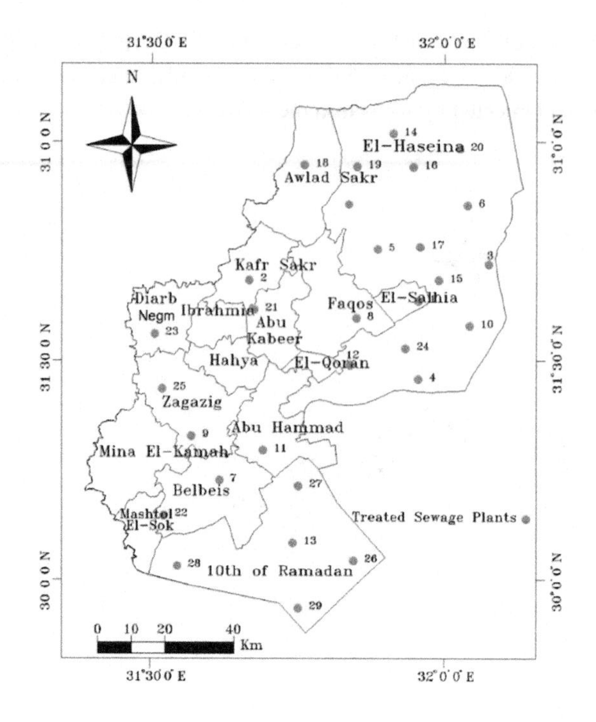

Figure 5.4. The sewage treatment plants in the Sharkeya

2 Extraction of brackish water

Desalination became very common lately in Egypt. It is considered as a viable option in water scarce regions. The main target of proposing this adaptation measure is to benefit from the brackish water and transfer it to a freshwater resource. The choice of extraction of brackish water rather than saline water from the NDA is to minimize the cost of desalination and environmental implications. Moreover, the effluent of the desalination unit can be safely released into the sea, as the residual water salinity will be almost with the same range of the Mediterranean Sea. So, this approach requires less energy and it is more beneficial with respect to disposal of the treatment effluent - the brine. Usage of pumped and treated brackish water has already been implemented nowadays in Egypt with reasonable cost, but large extraction of brackish groundwater from coastal aquifers has not been practiced [30].

As long as the extracted groundwater has concentration below 17.5 kg/m^3, (50% of the salinity of seawater), desalination treatment can produce fresh water quantities, which are about 50% of the brackish water inflow, with brine concentration below 30 kg/m^3, which can safely be discharged into the sea without affecting the marine environment. The main gain of this measure is the extracted groundwater, which can directly be used after desalination treatment.

For the case of the Sharkeya governorate, several trials were performed. One option was selected to be presented here with 274 extraction wells located at different depths in the brackish zone, with extraction of 2000 m^3/day each, pumping a total of about $200x10^6$ m^3/year of brackish groundwater, or about $18x10^9$ m^3 in the considered period of 90 years (Figure, 5.5). Assuming 50% efficiency after desalination treatment, this would result in about $9x10^9$ m^3 additional fresh water available over the same period.

A desalination unit could be built in the Sharkeya governorate. The main objective of the desalination plant is to desalinate brackish water not saline water. This will lessen the overall cost and the negative environmental impacts of production of concentrated brine from desalination of saline water. The output salinity concentration of the effluent after desalination process will be less in case of using brackish water. The overall cost including maintenance per 1 m^3 of desalinated water in Egypt is 4-6 EGP (equal approximately 20-30 cent in 2020) using vapour compression method. Evaluation of desalination and water transport costs is available in [31].

Changing of cropping patterns and irrigation practices

Cultivating crops with low irrigation requirements and more efficient irrigation practice could be a good adaptation measure. Such measure has been promoted and implemented in many water scarce regions [32]. It can be a cost-effective measure with large savings of fresh water, but it is associated with changing farming practices that have to be carried out over long periods. Combining incentives for farmers with effective education and training programs is required to realize the required changing of cropping patterns and irrigation practices [33]. This measure can result in usage of smaller amounts of water for irrigation, which will have consequently impact groundwater availability in the NDA (change of extraction and recharge). The assumption is that the increase in irrigation efficiency translates to real water savings, and additional water resources are not use for alternative uses (e.g., increase of irrigated area, increase of cropping density, non-agricultural uses etc.). Shifting to maize cultivation with lower water requirments instead of rice could reduce water for irrigation by 50% [34].

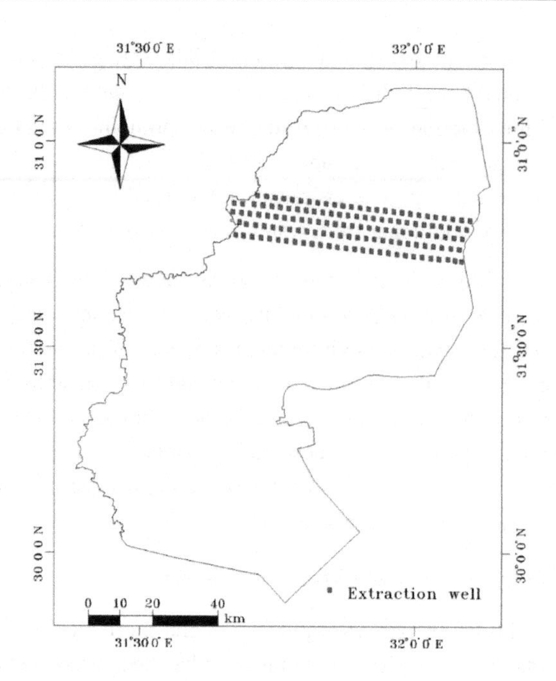

Figure 5.5. The distribution scheme of the extraction wells in the Sharkeya

For the Sharkeya governorate, this adaptation measure was tested by reducing the recharge by 50%, as consequence of implementing this measure, and simultaneous reduction of groundwater extraction from existing wells by 50%. This approach is very difficult to apply as it requires strict monitoring of the farmers and introduction of a concrete agricultural plan that the farmers follow. An intensive cooperation between (local) authorities and water users is essential to control the extraction. Beside investment in infrasture, educating, training, informing the water users and participation of water users in regular decisions could be very effective in achieving lower water use. In addition, groundwater extraction could be restricted through a system of permits.

5.5 RESULTS AND DISCUSSIONS

5.5.1 Injection wells

The application of injection wells indicates that there is a marked increase in fresh groundwater volume. The hydraulic head around the wells has increased. It should be noted that the aquifer gains about 8.5 $x10^9$ m^3 with this measure (Table 5.1). This increase is attributed to the fact that the increased volumes of fresh groundwater in the aquifer push the brackish and saline groundwater out and does not allow as much intrusion as would be occurring without that measure. Of course, if that available groundwater is further used for additional extraction these conditions would also change. One of the key issues in getting such results is the continuous maintenance and strict control to ensure sustainability over long time. Figure 5.6 shows the difference between the situation in 2100 with and without the adaptation measure (well injection) and the hydraulic head in both cases.

The use of the well injection measure to control SWI is an expensive measure with high operation and maintenance costs. Some disadvantages have also been reported including well clogging that causes a reduction of permeability around the well [8]. A serious investment in understanding the system is required (much better characterization of the geological conditions, estimation of the fresh-brackish-saline distribution), while an intensive, monitoring is required too.

Despite these drawbacks, this adaptation measure is considered to be a viable solution and has long been practiced worldwide (especially in Israel) to control SWI. This adaptation measure is feasible due to the availability of external water resources. The increase of freshwater and light brackish water keeps injection wells as a promising solution.

5.5.2 Extraction of brackish water

Regarding the extraction of brackish water adaptation measure, the results show that the brackish water zone has been reduced intensively due to groundwater extraction as depicted in Figure 5.7. However, it provides additional $9x10^9$ m^3 after desalination, which is ready for usage (Table 5.1).

The readily available water for use from the desalination plants and the fact that no extra water is required are the advantages of this measure. However, it is also a measure that requires investment in desalination plants and distribution network and an appropriate continuous monitoring network of the aquifer conditions to check the status of water type (brackishness within certain limits). Furthermore, additional analyses are certainly required regarding spatial spreading of saline groundwater and the general depletion of the NDA. Monitoring groundwater quality on a regular continuous basis is important. A network of observation wells around the brackish extraction well groups should be carried out and groundwater samples analysed. These well groups should be located at different depths and at such a distance that in case of salinization the progress of the saline water can be detected in appropriate time. In deeper layers where saline water is present, observation wells will also be required, below the extraction wells, and deep enough to detect early upconing.

5.5.3 Changing of cropping patterns and irrigation practices

For the third adaptation measure, according to Figure 5.8, the decrease in the recharge rate does not bring significant changes inside the aquifer in terms of available fresh groundwater in 2100 (gain of about $0.5x10^9$ m^3) compared to the most likely scenario. Simultaneous changes in both recharge and extraction may lead to limited changes inside the aquifer in terms of fresh/brackish/saline groundwater distribution, but the overall savings in terms of fresh groundwater can be quite significant. However, the amounts of water saved under the assumptions of 50% reduction of irrigation water and consequent recharge, in combination with 50% reduction of groundwater extraction from the wells for irrigation, are quite significant over the same period of 90 years. Therefore, considering these savings, it seems that this adaptation measure may be promising. Note that these savings in fact do not include additional reductions in intakes of surface water for irrigation.

However, as mentioned earlier, implementation of this measure on large scale (here assumed to take place across the whole governorate) is a challenging process that may take a long period. It should also be noted that the water quality of the desalinized water in the extraction of brackish water adaptation measure is higher than the irrigation water. These considerations, together with other implementation-related aspects need to be taken into account for the final overall choices regarding best adaptation measures.

These initial results demonstrate that with the considered adaptation measures, the conditions of the NDA in the Sharkeya governorate can be improved. Further analyses need to take into account combination of measures and more detailed (agro-economic) analyses of the suitability of different locations for implementing different measures. These further analyses can be performed with the developed model (possibly using a finer grid resolution locally to better reproduce saline groundwater upconing processes).

Table 5.1 shows the comparison among the three proposed adaptation measures for the likely future scenario with respect to the available water the NDA, the loss/gain of freshwater in the aquifer and the additional freshwater available over a period of 90 years. These results indicate that in terms of fresh groundwater inside the aquifer the best results are found with well injection adaptation measure, followed brackish water extraction and changing of cropping patterns and irrigation practices. However, changing of cropping patterns and irrigation practices gives the highest gain in total available freshwater.

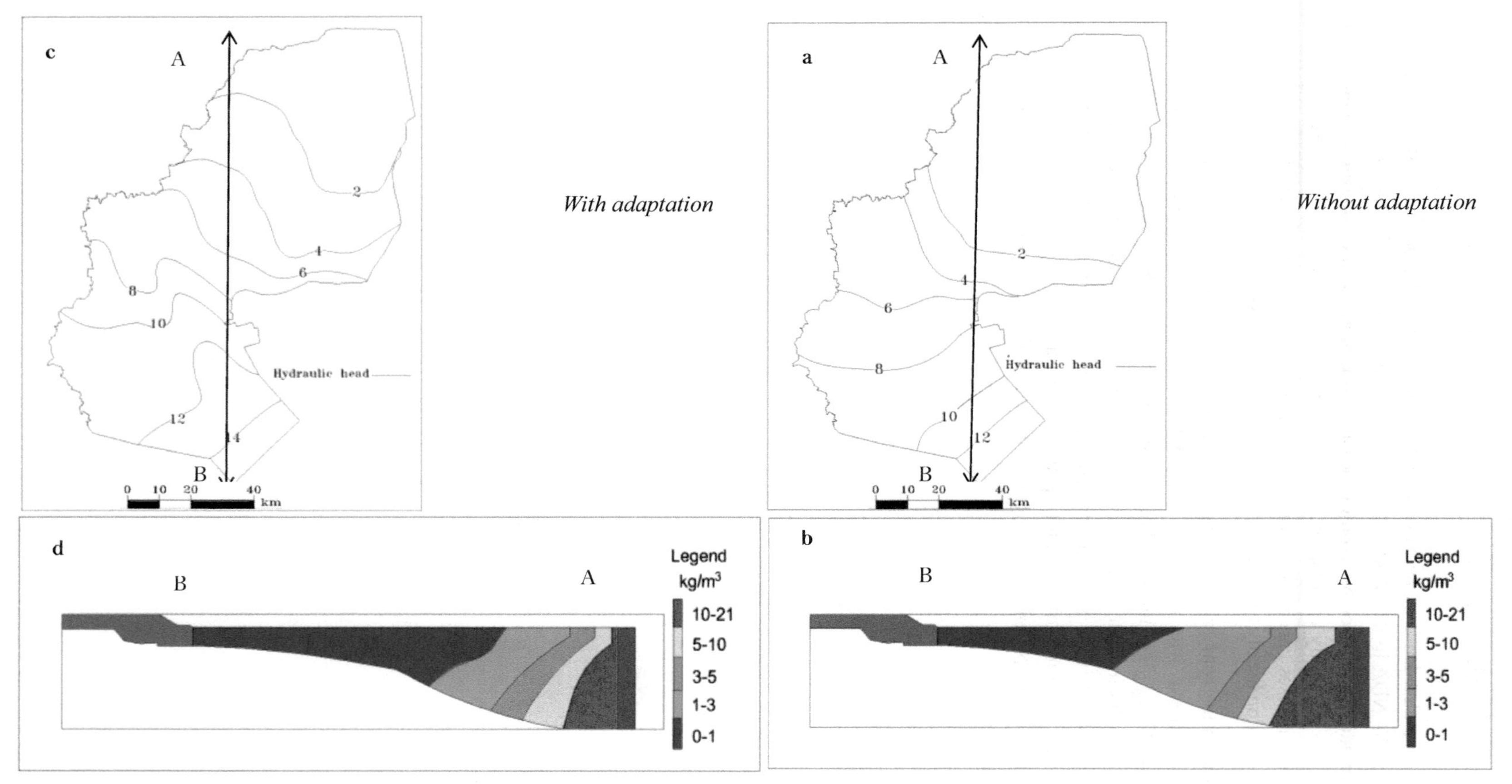

Figure 5.6. a: Hydraulic head without adaptation b: The cross section A-B in 2100 without adaptation. c: Hydraulic head with well injection d: The cross section A-B in 2100 with well injection(the vertical scale 1cm:400m)

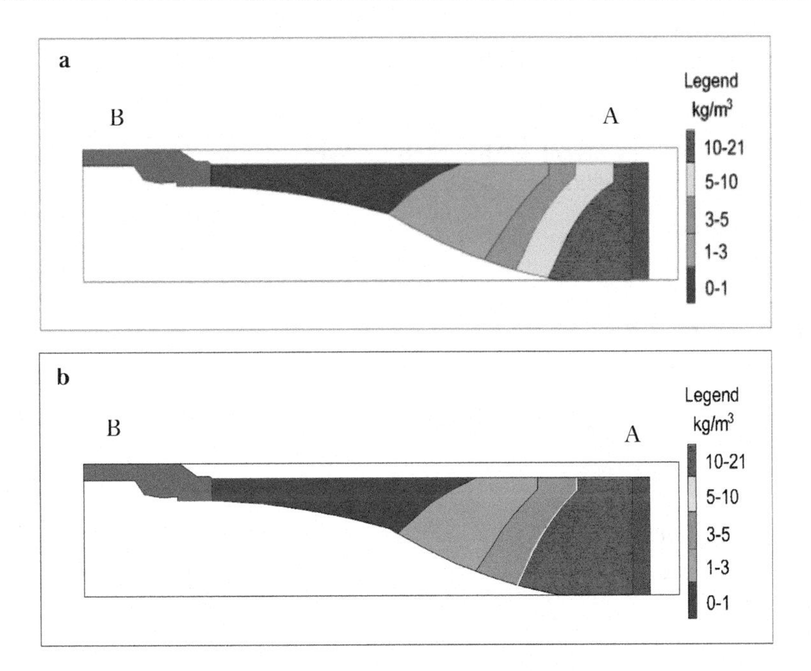

Figure 5.7. a: Cross section A-B without adaptation measure. b: Cross section A-B in case of extraction of brackish in 2100(the vertical scale 1cm:400m)

Table 5.1. Comparative results for the three proposed adaptation measures in case of the Sharkeya governorate

Parameters / conditions assessed	Current conditions 2010	Most likely scenario 2100 [2]	Well injection 2100	Brackish groundwater extraction 2100	Changing cropping patterns & irrigation practices 2100
Available fresh water in the aquifer in the given year in 10^9 m^3	322	295	303.5	296	295.5
Loss/gain of freshwater in the aquifer in 10^9 m^3	0	-27	-18.5	-26	-26.5
Additional freshwater available over period of 90 years (2010-2100)	-	-	-	9	10.5
Total loss of freshwater (2010 2100)	-	-27	-18.5	-17	-16

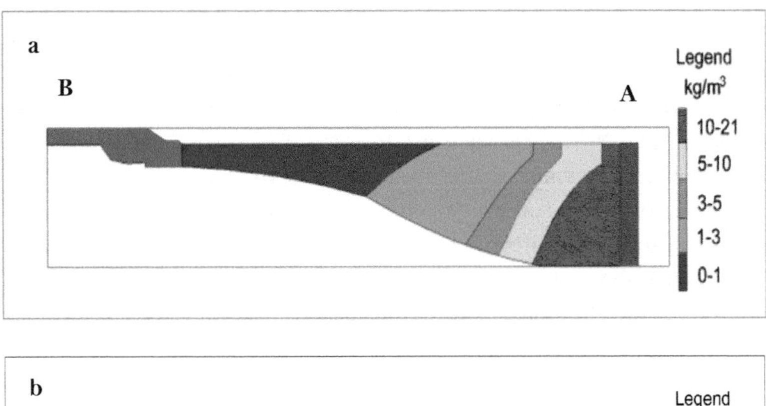

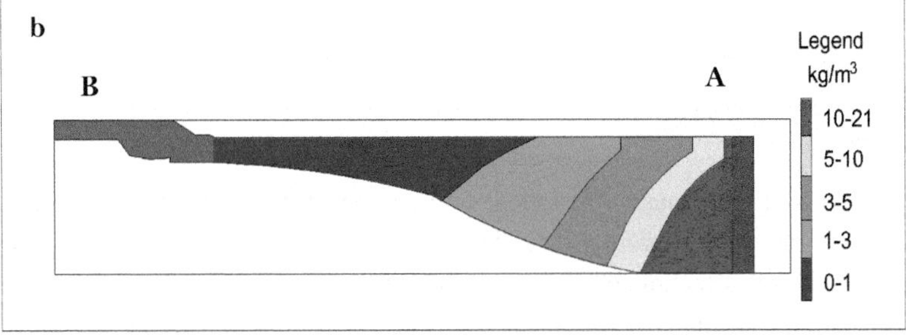

Figure 5.8.a: *Cross section A-B without adaptation measure b: Cross section A-B in case of changing of cropping patterns and irrigation practices in 2100 (the vertical scale 1cm:400m)*

5.6 CONCLUSIONS AND RECOMMENDATIONS

SLR and severe groundwater extractions are very likely to cause significantly increasing salinization impacts in the NDA. To assess current and future conditions of the NDA, a 3D regional variable-density groundwater flow model was developed using the computer code SEAWAT. The representative model for year 2010 was used as a predictive tool for assessing future fresh groundwater resources under SLR and extraction scenarios. The Sharkeya governorate was chosen as a case study and three adaptations measures were proposed. The possible measures include injection wells, extraction of brackish water, and changing of cropping patterns and irrigation practices. The results of testing the three potential adaptation measures were presented. From freshwater quantity point of view, changing of cropping patterns and irrigation practices could to be a promising measure. However, the complexity of the problem requires likelythe implementation of a combination of different measures, accompanied with detailed analyses of best locations for particular measures and agro-economic considerations.

Most of the adaptation measures appear to be expensive and need strict monitoring. However, the value of mitigating the salinization hazard in the NDA before it is too late is invaluable. Besides technical adaptation measures, an intensive cooperation between governmental authorities and water users in the Sharkeya governorate is necessary to control groundwater extraction. Involving stakeholders in decision-making could be effective in managing water use. Minimizing agricultural activities, shifting to other governorates, a system of fines for violation of extraction caps and using salt tolerant crops could lessen groundwater extraction [22]. As agriculture is the main water-consuming sector, shifting to industrial or service sectore activities as alternatives would also decrease the pressure on water resources.

References

1 Oppenheimer, M., B.C. Glavovic, J. Hinkel, R. van de Wal, A.K. Magnan, A. Abd-Elgawad, R. Cai, M. Cifuentes-Jara, R.M. DeConto, T. Ghosh, J. Hay, F. Isla, B. Marzeion, B. Meyssignac, and Z. Sebesvari, Sea level rise and implications for low-lying islands, coasts and communities. In: IPCC special report on the ocean and cryosphere in a changing climate, 2019.

2 Mabrouk, M.; Jonoski, A.; Oude Essink, G.H.P.; Uhlenbrook, S. Impacts of sea level rise and groundwater extraction scenarios on fresh ground water resources in the Nile Delta Governorates, Egypt. *Water J.* 10, 1690, 2018.

3 Sestini, G. ND., A review of depositional environments and geological history. Geol. Soc. Lond. Spec. Publ., 41, 99–127, doi:10.1144/GSL.SP.1989.041.01.09, 1989.

4 Mabrouk, M.B.; Jonoski, A.; Solomatine, D.; Uhlenbrook, S. A review of seawater intrusion in the ND groundwater system—The basis for assessing impacts due to CC, SLR and water resources development. *Nile Water Sci. Eng. J.* 10, 46–61, ISSN 2090-0953, 2017.

5 Anwar, H.O. The effect of a subsurface barrier on the conservation of freshwater in coastal aquifers. *Water Research J.* 17, 10, 1257–1265, 1983.

6 Luyun, R., Momii, K., and Nakagawa, K.: Effects of recharge wells and flow barriers on seawater intrusion, *Ground Water J.* 49, 239–249, 2011.

7 Bruington, A.E., and F.D. Seares, Operating a seawater barrier project. *Irrigation and Drainage Eng. J.* 91, 1, 117–140, 1965.

8 Bear, J. Hydraulics of groundwater, Dover Publications, Courier corporation, 592, ISBN 0-486-45355-3, 2007.

9 Hunt, B., Some analytical solutions for seawater intrusion control with recharge wells.doi.org/10.1016/0022-1694 (85)90072-1, 1985.

10 Mahesha, A., Transient effect of battery of injection wells on seawater intrusion. *Hydraulic Eng. J.* 122, 5,266–271,1996a.

11 Mahesha, A., Steady-state effect of freshwater injection wells on seawater intrusion in coastal aquifers. *Irrigation and Drainage Eng. J.*122, 3, 149–154,1996b.

12 Tsanis, I.K., and L.-F. Song, Remediation of seawater intrusion: A case study. *Ground Water Monitoring and Remediation J.* 21, 3, 152–161, 2001.

13 Oude Essink, G. H. P. Improving fresh groundwater supply problems and solutions, *Ocean Coast. Manage. J.* 44, 429–449, 2001.

14 El Raey, M., Frihy, O., Nasr, S. M., and Dewidar, K. H., Vulnerability assessment of sea level rise over Port said governorate, Egypt, Kluwer Academic Publishers, *Environ. Monitor. Assess. J.* 56, 113–128, 1999.

15 El-Shaer, M. H., H. M. Eid, C. Rosenzweig, A. Iglesias, and D. Hillel. "Agricultural adaptation to climate change in Egypt." In adapting to climate change, 109-127. Springer, New York, NY, 1996.

16 Nofal, E.R., Fekry, A.F., El-Didy, S.M. Adaptation to the impact of sea level rise in the Nile Delta coastal zone, Egypt. Am. Sci. J., 10, 17–29, ISSN: 1545-1003, 2014.

17 Mabrouk, M., Jonoski, A., Oude Essink, G.H.P., Uhlenbrook, S. Impacts of sea level rise and groundwater extraction scenarios on fresh groundwater resources in the Nile Delta governorates, Egypt. *Water J.* 10, 1690, 2018.

18 CAPMAS. The Central Authority for Public Mobilization and Statistics, Egypt, Egypt in Numbers; Ministry of Communication and Information Technology (publishing center): Cairo, Egypt, 2019.

19 EHSW, National yearly report. Egyptian Holding Company for Sewage and Water Supply, 2018.

20 Van Engelen, J., Oude Essink, G.H.P., Kooi, H., Bierkens, M.F.P. On the origins of hypersaline groundwater in the NDA. *Hydrol. J.* 2018, 560, 301–317. doi:2018.03.029/hydrol10.1016, 2018.

21 Todd, D.K. Ground water hydrology. New York: John Wiley & Sons, 1959.

22 Van Dam, J.C., Exploitation, restoration and management. In seawater Intrusion in coastal aquifers—concepts, methods and practices, ed. J. Bear et al., 73–125. Dordrecht, The Netherlands: Kluwer Academic Publishers, 1999.

23 Yuan J., Van Dyke M.I. & Huck P.M., Water reuse through managed aquifer recharge (MAR): Assessment of regulations/guidelines and case studies. *Water Quality Research J.* 51, 4 357–376. doi: https://doi.org/10.2166/wqrjc.2016.022, 2016.

24 Sprenger, C., Hartog, N., Hernández, M., Vilanova, E., Grützmacher, G., Scheibler, F. & Hannappel, S., Inventory of managed aquifer recharge sites in Europe: historical development, current situation and perspectives. *Hydrogeol. J.* 25, 1909–1922. doi:10.1007/s10040-017-1554-8, 2017.

25 Dillon, P., Stuyfzand, P., Grischek, T., Lluria, M., Pyne, R.D.G., Jain, R.C., Bear, J., Schwarz, J., Wang, W., Fernandez, E., Stefan, C., Pettenati, M., van der Gun,

J., Sprenger, C., Massmann, G., Scanlon, B.R., Xanke, J., Jokela, P., Zheng, Y., Rossetto, R., Shamrukh, M., Pavelic, P., Murray, E., Ross, A., Bonilla Valverde, J.P., Palma Nava, A., Ansems, N., Posavec, K., Ha, K., Martin, R. and Sapiano, M., Sixty years of global progress in managed aquifer recharge. *Hydrogeology J.* 27, 1, 1–30 https://doi.org/10.1007/s10040-018-1841-z, 2019.

26 Zuurbier K., Smeets P., Roest K. & van Vierssen W., Use of wastewater in managed aquifer recharge for agricultural and drinking purposes: The Dutch experience, in safe use of wastewater in agriculture, H. Hettiarachchi and R. Ardakanian (eds.). ©UNU-FLORES. 159-175, https://doi.org/10.1007/978-3-319-74268-7_8, 2018.

27 Zekri, S., Ahmed M., Chaieb R. & Ghaffour N., Managed aquifer recharge using Quaternary-treated wastewater: An economic perspective. *Int. Water Res. Development J.* 30, 2, 246–61, 2014.

28 World Bank Beyond Scarcity: Water Security in the Middle East and North Africa, MENA Development Series. World Bank, Washington, DC. License: Creative Commons Attribution CC BY 3.0 IGO, 2018.

29 Ebrahim, G., Jonoski, A., Al-Maktoumi, A., Ahmed, M. & Mynett, A. Simulation optimization approach for evaluating the feasibility of managed aquifer recharge in the Samail lower catchment, Oman. *Water Res. Plan. and Manag. J.*, ASCE doi: 10.1061/ (ASCE) WR.1943-5452.0000588, 05015007, 2015.

30 Stofberg, S.F., Zuurbier, K.G., Janssen, G.M.C.M., Oude Essink, G.H.P., Van Baaren, E.S., Boonekamp, T., De Buck, W., Hulzebos, J., Schetters, M. & Zwolsman, G. COASTAR: Exploration of strategic brackish groundwater extraction (in Dutch: Verkenning strategische brakwaterwinning) Deltares, technical report 11200070-001-BGS-0002 & KWR2018.02, 2018.

31 Djebedjian B, Mohamed MS, El-Sarraf S, Rayan MA. Evaluation of desalination and water transport costs (case study: Abu Soma Bay, Egypt). Nine Int. water tech. Conf., Sharm El-Shiekh, Egypt, 2005.

32 Van Bakel, P.J.T., Kselik, R.A.L., Roest, C.W.J. & Smit, A.A.M.F.R. Review of crop salt tolerance in the Netherlands, Alterra-report 1926, ISSN 1566-7197, 2009.

33 El Gafy, I., Ramadan, N. Hassan, D., Economic irrigation water productivity maps for Egyptian governorates, official publication of the European water association (EWA), ISSN 1994-8549, 2014.

34 C. Brouwer and M. Heibloem, Irrigation water management, training manual no. 3, Part II determination of irrigation water needs, ch. 2, FAO, 1986.

6 CONCLUSIONS AND RECOMMENDATIONS

Water scarcity is one of the main challenges that Egypt faces. Population growth and unfulfilled development requirements intensify the extraction of groundwater to meet the deficiency in fresh surface water. On the other hand, SWI is a significant threat to coastal aquifers e.g. NDA. In recent years, groundwater quality in the ND shows an increase in salinity concentrations that exceeds acceptable limits. This degradation in groundwater quality is attributed to SWI and extensive extraction in the NDA causing a serious limitation to the utilization of that valuable resource and imposing critical challenges on agricultural sector. Nevertheless, the ND is considered the most fertile land in Egypt. These conditions stress the need to understand and analyze the hydrological conditions and their impact on groundwater behavior in the NDA.

This research contributes to clear understanding of the current situation of the NDA based on updated reliable data on groundwater salinization and concrete knowledge of the hydrological, hydrogeological, geological and hydrochemical characteristics of the NDA, which will enable the decision makers to manage this valuable but vulnerable resource properly.

A 3D model simulating regional variable density groundwater flow was carried out, using the SEAWAT code to assess the current situation of groundwater salinization in the NDA and develop future adaptation strategies. A method for the identification of the most representative model has been used, based on testing different simulation periods during which the NDA has 'evolved' from completely fresh groundwater conditions to conditions representative for the year 2010, for which most data on salty groundwater conditions were available.

This model was then applied to test the NDA conditions under several pre-defined scenarios of SLR and groundwater extraction. This analysis indicates that impacts from further extractions of groundwater on fresh groundwater availability in the NDA are more significant compared to those from SLR.

Three different adaptation measures and their impacts in the Sharkeya governorate were initially tested, indicating that changing of cropping patterns and irrigation practices to water saving options seems to be a promising measure; in terms of the quantity of fresh water, when compared to artificial recharge of the aquifer with injection wells or extraction and usage of brackish groundwater after desalination treatment.

In the following, the posed research questions in the beginning of this thesis (Chapter 1) are addressed based on the findings from this research.

1 What is the current knowledge regarding groundwater salinization in the study area, and where are the knowledge gaps?

As we have seen in chapter 2, CC and its impacts on the NDA were the subject of many comprehensive studies for the past 30 years. Recent years have brought scientific evidence that SLR and excessive extraction are affecting the quality of the groundwater in the NDA. Many researchers have studied these impacts from different perspectives and confirmed that significant groundwater salinization has affected the NDA and that it will rapidly become worse in the future.

Serious negative socioeconomic impacts can follow as a consequence. This situation prompts for the studying and analyzing of the problem thoroughly and identifying flexible adaptation measures that can not only mitigate the negative effects of CC, but also lead to capacity development for coping with uncertain future changes.

Previous research shows that there is a gap in the studies that focus on sustainable groundwater resources development and environmentally sound protection as an integrated regional plan in the ND. The reason that prevents scientists from advancing research in the ND is the lack of data of sufficient quality. Having a complete set of data series is especially problematic in Egypt.

For the NDA, most research on quantifying variable-density groundwater flow processes have been carried out using 2D models, which cannot capture the full dynamics of the fresh groundwater-SWI. The majority of reported modeling studies were of local nature, implemented in specific regions to analyze the problems of a particular zone and interpret the results in terms of impacts caused by local causes. However, NDA should be integrated together in order to identify their relations and influences. Recently, some studies have shown the applicability of variable-density groundwater flow modeling as a useful instrument. However, there is still a need to develop a reliable regional 3D model that can serve as a tool for analyzing future scenarios and potential adaptation measures, which the approach was taken in this research.

Moreover, although there is extensive extraction in the ND, little is known in the combined influences of SLR and development-related groundwater extractions. Special emphasis should be put on research dealing with critical conditions/combinations of SLR and groundwater extractions. In addition, few researchers have addressed identification of hotspots for salinization hazards in groundwater for some regions of the ND using modeling results. Existing CC scenarios could be used to formulate the possible future SLR and hydrological conditions, while development plans within Egypt could offer information for estimating future levels and spatial distribution of groundwater extractions.

Regarding adaptation and mitigation measures, the analysis of previous studies shows that very limited number of studies addressed these issues. There is also a need for that type of research linking numerical modeling results with socioeconomic constraints in an integrated approach.

2 **What is the current situation of salinization in the Nile Delta and its governorates? What are the recommended locations for extraction from the current perspective (2010)?**

A fully 3D variable-density groundwater flow model to simulate the current situation of salinization in the NDA as discussed in chapter 3 was developed to answer this research question. The model has been developed using a large data set available from different private and public organizations in Egypt.

An innovative procedure is performed. It starts from completely fresh ground water in the NDA and sets up several different simulations periods to obtain the best salinity concentration distribution that represents the year of 2010. After the simulation, the 3D fresh-saline distribution that best fits the observed salinity data is chosen. The developed 3D model represents the salinity distribution in the NDA well. The results clearly demonstrate that salinization patterns are extensively invading the groundwater resources of the NDA. The comparison between numerous spatially varying groundwater salinity concentrations observed and modeled results shows that the model is capable of representing the current salinity distribution in the NDA, taking the year 2010 as the reference year. Therefore, we believe that the model is as reliable as possible given data constraints, and it can be used with confidence for future predictions.

Regarding the results obtained at the end of the simulation period, which represent the situation in the year 2010, the saline groundwater has spread in the north and northeastern more widely than in the northwestern, possibly because of the presence of a geological formation with higher hydraulic conductivities. The salinity concentration varies from higher than 30 kg/m^3 to 10 kg/m^3 in the north. The groundwater in the southern and middle regions is virtually fresh, being far distant from SWI introduced at the Mediterranean Sea boundary. The salinity concentration in the southern region is very low, with values of less than 0.05 kg/m^3. The total volume of groundwater in the NDA up to the hydrogeological base was estimated to be approximately 4050×10^9 m^3.

The thesis discusses in detail the recommended locations in the ND for future extraction activities. The results show that the middle part of the Delta is less vulnerable to SWI than its fringes. Therefore, the government should prohibit extraction at both sides of the ND (eastern and western). The model results indicate that specific regions in the east (Sharkeya governorate) and southwest (El-Buhaira governorate) are likely to suffer from salinization due to both natural (geological) reasons, dissolution of marine deposits and man-made (excessive extraction of groundwater). Both governorates have great agricultural and economic value in Egypt, especially in the export of commodities. Further studies on the fresh water needs and methods of extraction in these governorates might impact both conservation efforts and economic development planning with a view to conservation of the fresh groundwater resources in the NDA.

3 **What is the impact of saltwater intrusion under the various proposed future scenarios of climate change (sea level rise) and development (groundwater extraction) in the whole Nile Delta aquifer?**

The usage of the developed model has been demonstrated by developing a set of six possible future scenarios considering further increase of groundwater extraction in the NDA and possible SLR in the Mediterranean Sea due to climate change. The groundwater salinity conditions were analysed by using the change in volume for the four-groundwater types in the whole ND and for each ND governorate separately. The total volume of groundwater in the NDA up to the hydrogeological base was estimated to be approximately $4050 x 10^9$ m^3

The first scenario runs until 2500 with no SLR or extra extraction. It shows that time alone can bring further SWI into the NDA. For scenarios 2 to 6, the models were run up until 2100 with different SLR rates and/or groundwater extraction rates. The results show that the in scenario 2, fresh water volume decreased significantly by 18.7% in 90 years, while the saline groundwater volume gain by approximately 12%. These values are the highest among scenarios 2 to 6, because Sc.2 shows the combined extreme effect of SLR and groundwater extraction rate. Sc. 3 (moderate) shows modest values for freshwater volume loss (−13.3%) and saline groundwater gain (+5.7%). Under very optimistic conditions, Sc.4 (Restrictive) has a small saline groundwater volume increase (+2.2%). The fresh groundwater loss (–7.7%) is the lowest among all other scenarios. To achieve this scenario conditions, very rigid control of groundwater extraction is necessary. In Sc.5 (high extraction), the freshwater volume decreases by about 15.5%. That decrease is the second in severity after Sc.2 (extreme) with approximately 18.7%. A decline in freshwater in Sc.6 (high SLR) reaches about 11%. This value is less than that in Sc.5 confirming that SLR has a smaller effect on the groundwater volumes in the whole NDA compared to groundwater extraction.

This thesis indicates that impacts from further extraction of groundwater on fresh groundwater availability in the aquifer are more significant compared to those from SLR. The thesis highlights that if uncontrolled development of groundwater resources continues while adaptation measures are not implemented, this valuable fresh groundwater resource will be impaired to an extent that negatively affects the overall socioeconomic development of the country. Therefore, management of groundwater in the ND and adaptation measures to salinization threats should become one of the top priorities in the Egyptian water agenda.

4 What are the best locations and the vulnerable ones for groundwater extraction in the Nile Delta governorates in a long-term perspective?

Scenario analysis per governorate that was discussed in chapter 4 answers this question. The most suitable locations for extraction in terms of salinity and permissible drawdown limits are the in middle regions of the NDA, as the aquifer is almost in steady conditions with no significant change under different proposed scenarios.

Coastal governorates (Kafr El Sheikh, Damietta and Alexandria) are very vulnerable to SLR. Damietta already has no fresh groundwater, while Kafr El Sheikh shows a severe reduction of fresh groundwater in all the future scenarios. It is recommended that groundwater extraction is banned in the northern coastal governorates that are very sensitive to SLR and groundwater extraction. No groundwater extraction or investment in agriculture should be made in the coastal governorates due to their shortage in fresh groundwater.

The fresh groundwater volume decreases at a lower rate in the southern governorates, as Qalyobeya compared to the coastal governorates. Under the Long Run scenario (2500), Qalyobeya governorate fresh groundwater volume will decrease from 66 to 51×10^9 m^3 compared to Kafr el Sheikh fresh groundwater volume that will be diminished.

Meanwhile, the fresh groundwater volume decreases at a high rate in the Sharkeya governorate which suffers from the combined effect of excessive groundwater extraction and SLR. The ground water volume under the Long Run scenario will intensively decrease from 318 to 210×10^9 m^3. The analysis of the spatial distribution of salinity indicates that in this governorate, the northern parts are already in a critical condition, while some continued groundwater extraction can still be allowed in the southern parts. Ismailiya governorate shows a significant drop of the (relatively small) fresh groundwater volume in all different scenarios. The groundwater volume under the Long Run scenario is predicted to decrease from 31 to 8×10^9 m^3.

In the southern governorates, like Qalyobeya and Monofeya, the groundwater extraction of fresh water can continue, as the results indicate that these governorates have only fresh and light brackish water and do not have any brackish or saline water in all future scenarios. Groundwater extraction can be allowed only in the southern parts of the Sharkeya and El Buhaira governorates, because their northern parts are already in a critical condition.

For strategic planning of groundwater management in Egypt until 2100, it is recommended that future groundwater extraction distribution is adjusted. The vulnerability of governorates in terms of the available fresh groundwater volume (under stress) should be taken into consideration when designing future water management adaptation measures.

5 What are the proposed adaptation measures that minimize the loss of fresh groundwater due to saltwater intrusion, and what are their limitations?

The thesis proposes three adaptation measures: well injection, extraction of brackish water and changing of cropping patterns and irrigation practices, which were tested with the model in the Sharkeya governorates as a study case.

The increase of freshwater and light brackish water keeps injection wells as a promising solution. The aquifer gains about $8.5x10^9$ m^3 with this measure. One of the key issues in getting such results is the continuous maintenance and strict control to ensure sustainability over time. One of the advantages of this measure is that this practice has been implemented worldwide to control SWI.

For the second adaptation measure; extraction of brackish water, the loss of fresh groundwater in the aquifer is $26x10^9$ m^3 instead of $27x10^9$ m^3 without adaptation. However, it provides an additional $9x10^9$ m^3 after desalination, which has a very high quality and does not have to be treated. The water from the desalination plants is readily available for use and the fact no extra water would be required are the advantages of this measure. However, it requires investment in desalination plants and a continuous monitoring of the aquifer conditions.

For the third adaptation measure, the decrease in the recharge rate does not bring significant changes inside the aquifer in terms of available fresh groundwater in 2100 (gain of about $0.5x10^9$ m^3) compared to the most likely scenario. However, the amounts of water saved under the assumptions of 50% reduction of irrigation water and consequent recharge, in combination with assumed 50% reduction of groundwater extraction from the wells for irrigation, are quite significant over the same period of 90 years. Therefore, considering these predicted savings, it seems that this adaptation measure may be the most promising measure from a fresh water quantity point of view, when compared to the other two adaptation measures.

However, as mentioned earlier, implementation of this measure on a large scale (here assumed to take place across the whole governorate) is a challenging process that may take a long time and require investments (more efficient irrigation technology) and capacity building. It should also be noted, that the water quality of the desalinized water in the extraction of brackish water adaptation measure is higher than the irrigation water.

The complexity of the problem likely requires the implementation of a combination of such measures, with further analyses of appropriate locations for particular measure, including agro-economic and social considerations.

Concluding remarks and general recommendation

In conclusion, the developed model can serve as a useful tool for further studies within the water-food nexus configuration in the ND. It could serve as a tool for assessing areas of the groundwater system vulnerable to salinization due to combined stresses, such as increased groundwater extraction and SLR. It can be used to track the movement of fresh, brackish and saline groundwater in the NDA, and for testing future adaptation measures for the NDA. It could also be used and tested in aquifer systems in various deltas around the world where groundwater resources are deteriorating due to SWI. In spite of differences in geometry and their hydrological data, most deltaic areas face similar development and climate stresses. Such significant use of the model, however, critically depends on data availability.

This thesis recommends further research and analysis for various model applications in different governorates of the ND (small scale) to study the impact of extraction on the depletion of groundwater resources accompanied by a detailed adaptation plan for groundwater salinization for each governorate. It should be noted, that socioeconomic studies for the impact of SLR in the ND are lacking although they are considered very important.

List of Acronyms

AMSL	Above mean sea level
CC	Climate change
Ext.	Extraction
Mod.	Modeled value
NDA	Nile Delta Aquifer
ND	Nile Delta
Obs	Observed value
SWI	Saltwater intrusion
SLR	Sea level rise
3D	Three-dimensional
2D	Two-dimensional
Conc.	Concentration
RMSE	Root Mean Square Error
MWRI	Ministry of Water Resources and Irrigation in Egypt
Sc.1	Scenario 1
Sc.2	Scenario 2
Sc.3	Scenario 3
Sc.4	Scenario 4
Sc.5	Scenario 5
Sc.6	Scenario 6

List of Tables

LIST OF FIGURES

ABOUT THE AUTHOR

In 1998, Marmar Badr Mabrouk graduated from the Faculty of Engineering, Cairo University as a civil engineer. In 2004, she received her Master degree in Water Resources Management from Tanta University, Egypt. Her thesis was titled "Management of subsurface water in Fayoum governorate and its effect on soil behaviour" in which she studied recent theories and approaches in the field of groundwater management. The thesis focused on analysis of water budget of the Quaternary deposits to estimate the quantity of storage and its impact on the groundwater levels in Fayoum governorate.

In 2003, she obtained a Professional Diploma in Information Technology at the American University in Egypt. She has taken a number of training courses, which cover several fields, e.g. guidelines for good practice of water policy, Nile decision support tool program, preparing demand scenarios for agricultural water in Nile Basin for 2030, negotiation skills and conflict resolution, results oriented monitoring and evaluation for international projects.

Marmar has joined several private environmental companies before she started to serve as a technical engineer in 2001 in Nile Water Sector (Ministry of Water Resources and Irrigation, Environment and Water Resources Management Department). Her main responsibilities included but are not limited to: hydrological data processing, river Nile forecasting, analysing the hydrological and meteorological data of the Nile basin. She has worked on a wide range of water management issues and has worked in multidisciplinary research projects. In 2006, she worked as a program specialist in the National NBI Office *(NBI – Nile Basin Initiative – is a partnership initiated and led by the riparian states of the Nile River to develop the river in a cooperative manner).* Her main role was to provide overall coordination and support to the Integrated Water Resources Planning and Management Project. For this, she worked on studies regarding hydrological and participatory aspects of water resources management in the Nile basin throughout its components; water policy, DSS and water planning & management. From 2001-2007, she also worked in the capacity building for Nile Basin Water Resources Management Project financed from FAO as a national database expert.

In 2010, she was awarded a Ford Foundation fellowship to undertake her Ph.D. studies at IHE Delft, the Netherlands, on the issues of climate change and development impacts on groundwater behaviour in the ND.

Publications during her Ph.D.

- Mabrouk, M.; Jonoski, A.; Oude Essink, G.H.P.; Uhlenbrook, S. A Review of seawater intrusion in the Nile Delta groundwater system - The basis for assessing impacts due to climate changes, SLR and water resources development, Nile Water and Engineering journal 2017, 1.
- Mabrouk, M.; Jonoski, A.; Oude Essink, G.H.P.; Uhlenbrook, S. Impacts of Sea Level Rise and Groundwater Extraction Scenarios on Fresh Groundwater Resources in the Nile Delta Governorates, Egypt. Water journal, 10, 1690, 2018.
- Mabrouk, M.; Jonoski, A.; Oude Essink, G.H.P.; Regional groundwater modelling for determining adaptation strategies for the Nile Delta Aquifer" IAHR congress 2019.
- Mabrouk, M.; Jonoski, A.; Oude Essink, G.H.P.; Uhlenbrook, S. Assessing the Fresh–Saline Groundwater Distribution in the Nile Delta Aquifer Using a 3D Variable-Density Groundwater Flow Model. Water journal, 11, 1946, 2019.
- Mabrouk, M.; Jonoski, A.; Oude Essink, G.H.P.; Uhlenbrook, S. Assessing the adaptation measures to impacts from salnization threats in the Nile Delta Aquifer. Submitted to Water journal 2020 (in review).

Netherlands Research School for the
Socio-Economic and Natural Sciences of the Environment

D I P L O M A

for specialised PhD training

The Netherlands research school for the
Socio-Economic and Natural Sciences of the Environment
(SENSE) declares that

Marmar Badr Mohamed Ahmed

born on 22 July 1974 in Cairo, Egypt

has successfully fulfilled all requirements of the
educational PhD programme of SENSE.

Delft, 6 November 2020

The Chairman of the SENSE board

Prof. dr. Martin Wassen

the SENSE Director of Education

Dr. Ad van Dommelen

The SENSE Research School has been accredited by the Royal Netherlands Academy of Arts and Sciences (KNAW)

K O N I N K L I J K E   N E D E R L A N D S E
A K A D E M I E   V A N   W E T E N S C H A P P E N

The SENSE Research School declares that **Marmar Badr Mohamed Ahmed** has successfully fulfilled all requirements of the educational PhD programme of SENSE with a work load of 33.8 EC, including the following activities:

SENSE PhD Courses

- Environmental research in context (2010)
- Research in context activity: 'Organizing Stakeholder Workshop on: Climate Induced Changes on the Hydrology of Nile Delta, Reducing Uncertainty and Quantifying Riks (11 January 2011 – Cairo, Egypt)'

Other PhD and Advanced MSc Courses

- Applied groundwater modelling, IHE Delft (2010)
- Endnote course, IHE Delft (2010)
- Academic writing. American University in Egypt (2010)
- Presentation skills, British council (2010)

Stakeholder interaction and activities

- Farmer meetings on soil salinity in the Nile with delta governorate and water associations, National climate change and risk mitigation institute, Egypt (2011)
- Workshop 'Groundwater planning and management under climate change' with managers of groundwater sectors discussing data availability and monitoring, Groundwater Research Centre & Ministry of Water Resources and Irrigation, Egypt (2001)
- 1st Prima stakeholders forum, The Egyptian Ministry of Higher Education and Scientific Research, Egypt (2017)

Management and Didactic Skills Training

- Organising workshop 'Climate Induced Changes on the Hydrology of Nile delta, Reducing Uncertainty and Quantifying Risk", The Academy of Scientific Research and Technology and Zagazig University (2013).

Oral and Poster Presentation

- *Regional groundwater modelling for determining adaptation strategies for the Nile Delta Aquifer.* The 38th IAHR World congress, 1-6 September 2019, Panama city, Panama
- *Climate Change and Development Impacts on Groundwater Resources in the Nile Delta, Egypt.* Deltas in Times of Climate Change, 29 September – 1 October 2010, Rotterdam, The Netherlands
- *Nile Delta at risk*, 47th Cairo Climate talk, 2 May 2013, Cairo, Egypt

SENSE coordinator PhD education

Dr. ir. Peter Vermeulen